# Amiga Desktop Video Guide

Guy Wright

**Published by**
**Abacus**

Second Printing    1990
Printed in U.S.A.
Copyright © 1989, 1990         Abacus
                                    5370 52nd Street SE
                                    Grand Rapids, MI 49512

Copyright © 1989, 1990         Guy Wright

This book is copyrighted. No part of this book may be reproduced, stored in a retrieval system, or transmitted in any form or by any means, electronic, mechanical, photocopying, recording or otherwise without the prior written permission of Abacus or Guy Wright.

Every effort has been made to ensure complete and accurate information concerning the material presented in this book. However, Abacus can neither guarantee nor be held legally responsible for any mistakes in printing or faulty instructions contained in this book. The authors always appreciate receiving notice of any errors or misprints.

Throughout this book, the author and editors cite trade names and trademarks of some companies and products. These citations neither imply endorsement of these products or companies, nor any affiliation between these products or companies and Abacus. The Appendices of this book provide a complete listing of these products.

```
Library   of   Congress   Cataloging-in-Publication   Data

Wright, Guy, 1953-
    Amiga desktop video guide / Guy Wright.
       p.   cm.
    ISBN 1-55755-057-3 : $19.95
1. Video recordings--Production and direction--Data processing.
2. Video tape recorders and recording.   3. Amiga (Computer)
I. Title
TK6655.V5W75    1989
621.388'33'0285416--dc20                                 89-26786
                                                              CIP
```

# Introduction

In the past few years video and computer technologies have advanced in their own directions and at the same time they have converged. Video cameras and equipment have been making use of computer technologies and computers like the Amiga have been getting closer and closer to video standards. Twenty years ago it was rare to find a computer in someone's home and just as rare to find video equipment there. Now, it's not uncommon for people to own both.

If you own an Amiga and a VCR you have the beginnings of a video production studio. Already you have more power than full-blown broadcast stations had a few decades ago. With a few extra pieces of equipment you can blend the power of the computer with your VCR to produce professional results. With the right software you can produce titles, graphics, even sophisticated computer-generated animations.

It was once said that the personal computer was a solution looking for a problem. If you could find something that a computer could do better than a pencil and piece of paper then you could make a fortune. It happened with the spreadsheet, word processor and database. Computers could also play certain kinds of games, help musicians and artists, and you just can't do telecommunications without a computer. Since computers are naturals at manipulating text someone came up with the idea of using a computer to generate professional quality printed documents. That was the start of desktop publishing.

From the beginning, people in the Amiga community have been told that the Amiga is the perfect computer for video. It generates a nearly perfect video signal, has graphic capabilities far beyond other PCs and it is inexpensive. The term Desktop Video was coined pretty much as marketing hype by people trying to capitalize on the Desktop Publishing boom. But even if the term came from some PR executive, it also communicates an idea. That anyone can use the Amiga for a multitude of video purposes.

If you are a computer person, then you know that you don't have to be an assembly language programmer to get results. There are a handful of basics that you need to learn and once you have those you can do amazing things with a computer. The same is true for video. You don't need a waveform monitor and vectorscope to shoot a birthday party but to get good results you need to understand a few things about video. To make desktop video a reality you need some understanding of both technologies. You don't need to be an engineer or a programmer. You don't need $50,000 of specialized equipment. All you need is some patience, creativity and common sense.

While the concept of using the Amiga for video purposes was (and still is) intriguing, very few people have done it successfully. The reasons are, in my opinion, that people who know about computers don't know about video and people who know about video don't know about computers. Both areas are filled with confusing acronyms, conflicting ideas and a simple lack of information.

If there was just one place where a person could go to find out about using both technologies. A single source of information that would explain in simple terms what you did and didn't need. If you could buy a book that would lead you step-by-step through all the confusion then for you desktop video might become a reality. Well, this is the book.

Guy Wright

# Acknowledgements

I wish to thank the following people for their assistance in putting this book together. All the folks at Abacus (particularly Jim Oldfield, Jim D'Haem and Arnie Lee), Peter Utz (although I have never met him), Peter Lullemann, Mathew Leeds, Philo T. Farnsworth, Paul Laursen and family (for nuttin' in particular), most of the folks at AmigaWorld, Amiga Plus and Amigo Times, System Eyes Computer Store, all the manufacturers who lent me equipment and software, all the people who answered questions, my friends and relatives who said things like "Oh, a book? That's nice" and "What is desktop video, anyway?", my kids Ariel and Scott who let me use the computer when they had important games to play and my wife Susan who understood, encouraged, tolerated and kept the bed warm until all hours.

# Table of Contents

Introduction .................................................................................iii

1. Terms and Definitions.......................................................... 1
   Desktop Video.................................................................. 3
   Broadcast Standards........................................................ 4
   Video Cameras and Camcorders..................................... 4
   Video Signal..................................................................... 5
   Recording and Playback Machines ................................. 7
   Combining Video Equipment........................................... 7
   Video Production and Special Effects ............................ 8
   Graphics............................................................................ 9
   Video Signal Monitoring and Processing......................10
   Video Editing...................................................................11
   Audio ..............................................................................13
   Lighting...........................................................................14
   Summary .........................................................................14

2. Camcorders and VCRs ........................................................15
   Professional Video .........................................................17
   Consumer Video .............................................................19
   Summary .........................................................................39

3. Genlocks................................................................................41
   The Three Amigas...........................................................43
   Encoders, Sync Generators and Genlocks ....................45
   Not All Genlocks are Created Equal..............................48
   Other Genlocks...............................................................50
   Hooking up a Genlock....................................................51
   Using the Genlock ..........................................................54
   Summary .........................................................................55

4. Digitizers and Scanners .....................................................57
   Why Digitize?..................................................................59
   Types of Digitizers..........................................................61
   Scanners..........................................................................62
   Selecting a Digitizer.......................................................63
   Hooking up a Digitizer ...................................................64
   Digitizing Process...........................................................67
   Summary .........................................................................68

5. Frame Grabbers - Frame Buffers ................................................. 69
   Real-time digitizers, Frame Grabbers and Frame Buffers ........... 72
   Selecting a Real-Time Digitizer ................................................. 74
   Selecting a Frame Grabber ....................................................... 74
   Selecting a Frame Buffer .......................................................... 76
   Connecting a Frame Grabber .................................................... 77
   Frame Grabber Tips .................................................................. 78
   Legal Issues ............................................................................... 80
   Summary .................................................................................... 82

6. Software, Paint and Image Processing Programs .................... 83
   Video Software .......................................................................... 85
   What You'll Need ...................................................................... 89
   The Basic Package .................................................................... 90
   Paint Programs .......................................................................... 91
      Choosing A Paint Program .................................................... 92
      Image Processors .................................................................. 94
   Paint and Image Processing Program Tips ............................... 95

7. Rendering, Ray Tracing and Animation Software ................... 101
   Structured Drawing - Rendering and Ray Tracing Programs ... 103
   Choosing a Rendering Program .............................................. 104
   Animation Software ................................................................ 108
   Types of Animation ................................................................. 108
   Choosing an Animation Program ............................................ 113
      Page Flipping Programs ...................................................... 113
      All-in-One Animation Programs ......................................... 115
   Rendering Package Tips .......................................................... 116
   Summary .................................................................................. 117

8. Video Titling .............................................................................. 119
   Text in Titling .......................................................................... 123
      Character, Line and Screen Options .................................. 124
      Character Options ............................................................... 124
      Line Options ........................................................................ 126
   Picking a Text/Screen Program .............................................. 128
   Choosing a Presentation Program .......................................... 130
   Choosing a Text/Screen/Display Program .............................. 132
   Titling Software Tips .............................................................. 134
   Summary .................................................................................. 136

9. Music and Videos ...................................................................... 137
   When to Add Extra Sounds ..................................................... 140
      Adding Sounds During Taping ............................................ 140
      Adding Sounds During Editing ........................................... 141
      Adding Sounds During Playback ........................................ 142
   Audio Equipment ..................................................................... 142
   Computer Audio ....................................................................... 146

   Music Software .................................................. 147
   Sound Samplers ............................................... 149
  Audio Tips ........................................................... 150
  Summary .............................................................. 151

10. Special Effects............................................................ 153
  Traditional F/X .................................................... 155
  Camcorder F/X .................................................... 156
  VCR F/X .............................................................. 160
  Genlocking F/X .................................................. 161
  Digitizer F/X ....................................................... 163
  Frame Grabber F/X ............................................ 165
  Paint Program F/X .............................................. 166
  Image Processor F/X .......................................... 170
  Rendering F/X .................................................... 171
  Animation F/X .................................................... 172
  Titling F/X .......................................................... 173
  Sound F/X ........................................................... 174
  Summary .............................................................. 176

11. Putting It All Together ............................................ 177
  System Configurations ....................................... 179
  Minimum Configuration .................................... 180
  Mid-Range Configuration .................................. 185
  Professional Configurations .............................. 189
  Step by Step ........................................................ 193
   Pre-Production ................................................. 194
   Production ........................................................ 200
   Post-Production ................................................ 201
  Summary .............................................................. 211

12. Advanced Techniques ............................................... 213
  Titling .................................................................. 216
  Educational Uses ................................................ 217
  Business Presentations ...................................... 218
  Demonstrations .................................................. 220
  Live Performances ............................................. 220
  Post-production Houses .................................... 221
  Summary .............................................................. 224

13. Conclusion ................................................................. 227

Appendices ........................................................................ 231
  A. Related Products ............................................. 233
  B. Magazines and Books .................................... 237
  C. Amiga Video Sources .................................... 241

Index .................................................................................. 251

# Chapter 1

## Terms and Definitions

# Chapter 1
# Terms and Definitions

One of the problems that most people encounter when they begin to get involved with computers or video is the terminology. Each field has a confusing array of words, terms, specialized equipment and acronyms. When you talk to an expert in either computers or video it only takes a moment before they switch into another language filled with arcane and mysterious phrases. It is easy to become discouraged even if you thought you knew about that area.

With both fields there are a lot of terms and highly specialized things that you really don't need to know unless you are planning to become a video engineer or systems programmer. But there are also a lot of things that you should know. I am going to assume that you know more about the computer side than the video side. That you know what a disk drive is, what RAM is, what color palettes are, etc. I will describe in more detail what some of the specialized computer software and hardware is as it is encountered but there are a lot of terms in both areas that should be defined up front.

## Desktop Video

The first term, and perhaps the broadest, is Desktop Video. What exactly does that mean? In the most basic sense it means using a personal computer to assist or enhance a video production. Computers have been used in video work for some time now but until recently they have been highly specialized, very expensive computers that only did one or two things. The early personal computers were capable of many things but combining computer output with video was very complicated. Computers just couldn't produce the right kinds of video signals. The newer personal computers like the Amiga were the first computers capable of producing a fairly clean video signal. When you add a device called a genlock the computer signal can be mixed with any other video signal as if the computer were just another video camera or VCR generating its own images. Since the Amiga is a very flexible machine with many graphic and sound capabilities it can do many jobs in video production.

## Broadcast Standards

There are three standards when it comes to broadcast television. NTSC, PAL and SECAM. In order they are:

**NTSC**  In 1953 the NTSC (National Television Standards Committee) presented the FCC (Federal Communications Commission) with an electronic standard for colors. Over the years NTSC has come to stand for many things, "broadcast quality" video, the American broadcast standard of 525 interlaced horizontal lines at 30 (well, actually it is 29.97 but who is counting?) frames per second (that is 262.5 horizontal lines at 60 fields per second on a television screen), 15.75 kHz (KiloHertz), and the entire broadcast video signal, not just a committee that established a color standard.

These days, when someone says that a signal is up to NTSC standards, they mean the entire signal. If a device or piece of software is designated an 'NTSC' version it usually means it is for use with American equipment (rather than European).

**PAL**  PAL (Phase Alternation Line) is the European Television Standard. It has more horizontal lines per screen (512) at a slower (50 fields per second) rate. Unless you are going to be moving to or from Europe the only thing you need to know right now is that PAL and NTSC equipment and software are NOT compatible. So if you live in Europe buy PAL equipment and PAL compatible software, if you live in the U.S. or Japan buy NTSC equipment and NTSC compatible software.

**SECAM**  SECAM (Systeme Electronique Couleur Avec Memoire) is the French Television Standard. It is unique to France, some African and Middle Eastern countries and communist block countries. Like PAL, the equipment and software are NOT compatible with NTSC.

## Video Cameras and Camcorders

The first step in video is turning a real world image into an electronic signal. You do this with either a video camera or camcorder.

**Camera**  Occasionally you will see a video camera offered at a ridiculously low price. Check the fine print and you will find that it is either a black and white camera or it has no recording capabilities, or both. In other words you have to have it connected to a video recorder. This is fine if you are doing nothing but studio work but portability is limited by the length of your cables.

**Camcorders**  Camcorders are video cameras with recorder functions built in. They are usually battery operated but most can use a power adapter (some can even use the power from a car with a lighter-powered adapter) Camcorders will be discussed in more detail in Chapter 2.

# Video Signal

A video camera or camcorder takes a real world image and turns it into a video signal. How they do this requires a fair bit of electronic wizardry best left for other books. The video signal produced has many parts all used for various purposes. There are basically two forms of video you might get from a camera, RGB or Composit. (Occasionally, you might see a video signal separated into Chroma and Luma.)

**RGB**  RGB (Red, Green, Blue) is the video equivalent of an artists pigment colors. With these three, a television can produce just about any other color. RGB indicates a **noncomposit** video signal where the three colors are separated (and have three separate cables). RGB produces a clearer picture, a benefit when using computers.

**Composit Video**  A television signal can be looked at as having three basic elements; **video, audio and sync.** When the video and sync are combined electronically the result is called composit video. Many times composit video will just be referred to as **video in** or **video out** on the back of a VCR or camera.

**Encoder**  An Encoder is a device that turns computer output (like the Amiga) from RGB to composit.

You probably won't need to know what each part is used for but here is a quick list of everything that can go into a broadcast video signal;

**Luminance**  Luminance is the black and white information in a video signal.

**Chrominance**  Chrominance is the color information in a video signal.

**Audio**  Audio refers to the sound. Sometimes stereo.

**Sync Signal**  Horizontal and vertical synchronizing signals, Vertical Interval, Vertical Sync Pulse, Vertical Interval Test Signal and Vertical Interval Reference Signal - All make up the sync signal used by the monitor or TV to produce a clear picture.

A broadcast signal also contains Experimental Non-program Information - For captioning, etc.

**Control track**  When talking about video on tape you will sometimes see references to a **control track**. This is a separate set of electronic pulses placed on the tape during recording. These pulses are used to synchronize the timing of the playback machine.

**Frames and fields**  In the U.S. a video picture is made up of 525 horizontal lines. A single picture is a **frame**, which changes 30 times a second. A frame is made up of two **fields** (262.5 horizontal lines at 60 per second). The first field in a frame 'draws' the odd numbered lines and the second field 'draws' the even numbered lines. This system of combining two fields results in an **interlaced** picture. The two interlaced fields make up a frame. The sync pulses, intervals, etc. tell the monitor when it has reached the end of a line, the end of a screen, the end of a field, etc.

When you start adding computers like the Amiga into this mix you must also talk about non-interlaced pictures and overscan. The Amiga gives you the option of displaying information in many resolutions. Two of your choices are interlaced or non-interlaced modes. Non-Interlaced mode lets you use more colors but is limited in resolution. Interlaced mode gives you higher resolution but you have fewer colors at your disposal and the monitor has a tendency to flicker. This is because some of the phosphors on the monitor screen begin to fade before the screen is redrawn and the slower scan rate is close to the level of a humans perception. There is a device available to correct this flickering on the Amiga.

**Overscan**  The other term is overscan. Most computers limit their display to an area in the center of a monitor screen leaving a colored border around the edges. Normal video signals will fill the entire screen (plus a bit more to accommodate the variety and sizes of television picture tubes). **Overscan** is when you have the Amiga use the entire screen.

**Radio Frequency**  Finally, when you combine all the elements of a video signal together (video, audio and sync) the result is **RF** (Radio Frequency). When a television station wants to broadcast a program they take an RF signal and modulate it so that it can only be picked up on a certain channel. Then they broadcast it. A television set first demodulates the signal, separates the RF components, sends the audio to the speaker, the video to the screen and uses the sync to make sure the picture is right.

## Recording and Playback Machines

In a security system or closed circuit video situation the image is sent straight from the camera to a monitor, but in most video applications you have to be able to save the video and audio information somehow and play it back later. You do this with a video recorder or video player. There are dozens of formats and dozens of brand names but here are the four basic types;

**VTR**  Video Tape Recorder. A machine that can both record and playback video tape. Originally VTR meant a reel-to-reel video tape machine.

**VTP**  Video Tape Player. A machine that can only play back video tape but not record.

**VCR**  Video Cassette Recorder. A machine that can both play back and record on video cassettes. These are your standard home video recorders. They come in all shapes, sizes, prices and formats. The most common VCRs are either VHS or Beta format but other formats have been developed recently each with their own pros and cons. VCRs will be discussed in Chapter 2.

**VCP**  Video Cassette Player. A machine that can only play back video cassettes.

## Combining Video Equipment

First, when you have more than one piece of equipment hooked together each piece is considered **up-stream** or **down-stream**. If you just had a camera feeding into a VCR, the camera would be considered up-stream and the VCR downstream. The simplest video set up might have just one camera, a VCR and a monitor (this is what a Camcorder is all in one unit) but to do more than just point, record and play back you need more.

Unfortunately, when you try to combine pieces of video equipment things get a little complicated. Cameras, VCRs and your computer all create their own sync signals and mixing syncs just doesn't work without extra equipment. There are a number of ways around the sync problem but they all mean providing just one sync signal to the recording unit. You can either use a sync generator or a genlock.

**Sync generator**  A sync generator is a device that produces a clean, stable electronic synchronizing signal, used by all your equipment, providing your equipment can use external sync signals. Unfortunately, most home equipment can't use external sync.

**Genlock**  A genlock is a device that lets you mix the video signals from more than one source by using the sync from an up-stream source. Basically, the genlock leaves the up-stream sync alone and only adds the video signal from the genlocked device. If you have a camera up-stream then a genlocked computer and then a VCR downstream, the VCR gets both the video and sync from the camera but only the video signal from the computer. The term 'Genlock' used as a noun makes some video perfectionists cringe because 'genlock' is something you do to equipment not the device that does it. Strickly speaking there are only 'genlocking devices', 'genlocking units' or 'genlockable equipment' however, the term 'genlock' has come to mean the device itself as well as the function it performs. Genlocks will be covered in more detail in Chapter 3.

## Video Production and Special Effects

So, once you get all your equipment working together you will want to do more than just record straight video from the up-stream equipment. You will want to do more than just switch from one source to another. You'll want to put titles, animations, fades, wipes, etc. into the final product. To do this requires even more equipment. Normally you would need a special device for each of these tasks but in desktop video the computer can handle many of these duties.

**Switcher Fader**  A Switcher, or sometimes a Switcher/Fader, lets you switch between video sources. A simple switcher is just a bank of buttons corresponding to the different video sources. Push a button and that video goes down-stream.

**Special Effects Generator**  A Special Effects Generator (SEG) is a device that lets you do wipes, fades and switches between video sources (providing they are 'sync'ed).

**Character Generator**  A Character Generator simply puts words on the screen (something computers do pretty well).

| | |
|---|---|
| **Chroma Key** | A Chroma Key, sometimes built into an SEG, lets you mix video signals by replacing a certain color from one source with video from another source. The weather forecaster standing in front of a satellite map is an example of chroma keying. Genlocks for the Amiga also work this way. |
| **Video Digitizer** | A Video Digitizer takes a video signal and turns it into bits and bytes that the computer can understand. The image can be stored, loaded into a paint program and manipulated by the computer. The two kinds of video digitizers available for the Amiga right now are real-time digitizers (that can grab an image almost instantly) and the kind that take a few moments to operate (these require a static image). |
| **Scanner** | A Scanner is a device that lets you capture a still, two-dimensional image and store it in the computer. They work in much the same way that a copy machine does. Most scanners will give you a much better image than a digitizer but they are limited to 2-D images and are fairly expensive. While their main application is in the desktop publishing field it is possible to use a scanner in place of a digitizer. |
| **Frame Grabber** | A Frame Grabber is a device that captures a single frame of video and stores it in a RAM buffer. This is how a digital freeze frame works. Some frame grabbers act like real-time video digitizers converting the image into bytes that the computer can manipulate. |

# Graphics

While graphics would normally fall under special effects there are so many forms of computer graphics in desktop video that it should have it's own section. The basic computer graphics are done with **Paint programs** that let you use your computer screen as an electronic canvas. The second level would be **image processing** software that lets you manipulate existing computer graphics. Next is **Rendering** and **Ray Tracing** software. These are programs that create the objects using pre-defined shapes or pure mathematics. You "describe" the objects you want to the computer and it draws them for you. It is much more complicated than that but the images can be spectacular.

| | |
|---|---|
| **Animation packages** | Next you have animation packages. Some of these will work with other paint programs and some of them have built-in paint programs of their own. Some use traditional animation techniques (like page flipping) while others have the computer calculate the movements of objects by themselves. |

When you get into computer animation there are two basic types; RAM animation and Single Frame animation. **RAM animation** is a complete scene that is played in real time by the computer and its duration is usually limited by the amount of RAM you have. **Single Frame** animation is a technique where a picture is drawn, saved to a single frame on tape, then the next picture is drawn, saved, etc. RAM animation can be a little easier and cheaper, but the durations are short. Single Frame animation has no time limit but it can be expensive. VCRs capable of accurate single frame recording are not cheap. Unless you have lots of money to devote to, equipment RAM animation is probably the way you will want to go.

## Video Signal Monitoring and Processing

If all video and electronic equipment was perfect desktop video would be much easier but it isn't. Even in the fanciest television studios there is a need for special equipment to monitor and clean up imperfect video. There are a number of specialized devices that you may eventually want to think about purchasing if you plan to do more than just family picnics and birthdays.

**Waveform monitor**  Perhaps the most important tool in the video studio is a waveform monitor. While the picture may look great on your screen, the only way that you can really see what is going on with your signal is with a waveform monitor.

**Vector Scope**  A Vector Scope is another specialized measuring device that will accurately show you how good or bad your colors are.

**Proc Amp**  A Proc Amp (processing amplifier) separates a video signal and replaces the old sync with new. At the same time adjustments can be made to the video as if it was coming straight from a camera. Proc Amps are not usually considered home equipment and therefore cost quite a bit.

**Time Base Corrector**  A Time Base Corrector (TBC) is a device that temporarily holds part of an incoming video signal and sends it back out in an ordered fashion. Since most VCRs are a little 'sloppy' when sending out signals sometimes a TBC can correct the timing and clean up the signal. Another advantage to a TBC is its ability to accept external sync letting you use a SEG to mix video signals from units with conflicting sync signals. Again, TBCs are fairly expensive.

**Color Correctors or Image Enhancers**

A Color Corrector or Image Enhancer is a device that corrects the color balance in a signal and/or cleans up video signals. There are a number of Color Correctors and Image Enhancers available for home use. Some of them perform like a Proc Amp while others just boost the entire signal. They range from about $50 for a simple filtering device that may or may not correct much, all the way up to thousands of dollars for a unit that lets you adjust white levels, contrast, tint, detail, even mix audio signals, etc. A very modest color corrector with image enhancement circuitry will cost you about $200 to $400.

# Video Editing

Video tape editing is more an art than a science. Unless you are doing live television (mistakes and all) you will want to do some video editing. There are a few devices and terms that are specific to video editing.

**Glitch**

When you stop and start a camcorder or VCR in the middle of recording and then play it back the monitor will freak out for a few seconds when it reaches that point. That is called a **glitch**. It happens because the control track has been interrupted and there is a momentary loss of sync between scenes or the VCR is trying to read two conflicting sync signals. The only way to avoid this is by using a camcorder or recorder that has Flying Erase Heads. **Flying erase heads** can give you clean, 'glitch' free edits.

The simplest editing configuration would be a camcorder that you stop and start when you wish to change scenes. The next simplest editing setup would be a camcorder and a VCR. If your original material is on the camcorder and you are moving scenes onto the VCR then the camcorder is referred to as the **slave deck** and the VCR as the **master deck**. Making a straight copy from one machine to another is called a **dub**. If your original material is in the slave deck and you copy it to the master deck that is one **generation**. If you copy the copy that is two generations, etc. Since every machine introduces a certain amount of error while playing back tape each generation is a little poorer than the last. On a VHS machine you can expect about 10% loss of quality with each generation, which means that the practical limit is about four generations.

You cannot edit video tape with scissors and scotch tape and expect even poor results (results are closer to terrible) so all editing is done electronically. There are two types of edits, assemble edits and insert edits. An **assemble edit** is when you keep adding new scenes to the end of a sequence (new video, audio and new sync too). An **insert edit** is when you replace one scene with another in the middle of a sequence (only replacing the video and audio but leaving the old sync on the tape). It is also possible to edit or not edit the sound during an insert edit.

**Editor controller**

Editing can be done manually by starting and stopping the master and slave by hand or you can get a device called an editor controller. An **editor controller** is a specialized remote control for the master and slave. Depending on it's features you can specify the edit points and it performs the edit for you. Like everything else in video, editor controllers vary in price and capabilities. They can be as simple as a single cable that connects two decks on up to frame accurate devices. Fortunately many camcorders and VCRs have inputs for remote operations. Unfortunately few of them are compatible with other manufacturers equipment. Some of the more common inputs are the Sony 5 Pin **Control L**, and **Control S** (which can be found on a number of newer camcorders and decks), Panasonic also has a 5 pin input and some other manufacturers use their own. There are even controllers that use infrared signals rather than cables. One of the exciting elements of desktop video is that there are ways to turn your computer into an editor controller no matter what brand of equipment you use.

Since **video heads** (the magnetic units in a VCR that pick up the signals off the tape) are mechanical devices they cannot get up to speed instantly. And since the heads need to be at speed when an edit occurs some editor controllers will also **pre-roll** the VCRs. If you wanted to assemble edit a new scene onto the master you would pick the start points on both the master and slave decks, the editor controller would stop both machines, rewind them a few seconds, start them both so the heads have time to get up to speed and then at the correct point would switch the master into assemble editing.

Finally, some very expensive editing equipment uses **SMPTE time coding** (Society of Motion Picture and Television Engineers, pronounced sim-tec). A SMPTE timing track, readable by the editor controller, is added to a tape giving you frame accurate editing capabilities.

# Audio

Even silent films had piano accompaniment in the theaters and a dead quiet video production would be, shall we say different. There is as much to audio as there is to video. At the simplest level you can just use the sounds picked up by the microphone on your camcorder. Narrating as you tape the scenes. Or you could put narration in later, this is referred to as **voice over**. You could use the computer to compose and playback your own musical score or add sound effects. Or you could combine elements and add pre-recorded music, voice over narration and computer generated sound effects to your tapes during the editing process.

**Specialized microphones**

While the microphones built into camcorders are suitable for most applications you may wish to purchase some specialized microphones. There are many kinds of microphones available but they fall into a few categories. **Low impedance** or **high impedance** (sometimes printed as LO Z or HI Z). All you need to know is that they are not interchangeable electronically. **Condenser**, **Dynamic** and **Crystal** are the three basic types of microphones and indicate the microphone's mechanics. **Omni-directional, Cardioid, Uni-directional** and **bi-directional** indicate in which directions the microphone works best. And there are microphones designed for specific applications (wireless, shotgun, lavalier, boom, even telephoto mikes that zoom with your camcorder lens).

**Mixers**

Whatever you do with audio you will probably want a mixer. A mixer lets you blend sounds or music from two or more sources. The simplest ones are fairly inexpensive and can do a pretty good job. That way you can mix sounds and music from your computer, stereo, CD player, audio tape decks, microphones and the sounds from the original tape.

**Audio digitizer**

Another piece of equipment you might want is an audio digitizer. These take audio information and store it in the computer in much the same way a video digitizer works. We will cover audio in more detail in Chapter 9.

It should be noted that creating music with a computer is a vast area complete with specialized terminology and equipment. We will discuss audio in more detail in Chapter 9

## Lighting

Perhaps one of the most important (and most neglected) areas of desktop video is proper lighting. If you have a camcorder and don't have at least one video light then go out and buy one now. While each new camcorder manufactured these days claims lower and lower **LUX** numbers (the minimum amount of light the camcorder needs) there is one simple yet vital rule. The better the lighting the better the results.

One problem people complain about in desktop video is the quality of their results. In many instances the problem can be traced back to poor lighting during the original taping. While the picture may look okay on a monitor the quality deteriorates faster than normal with each generation, losing 20%, or 30%, or more! Yes, camcorders can record an image in low light but the signal will be poor and it will just get worse during editing and dubbing.

## Summary

There are thousands of other terms and pieces of equipment in both the video and computer fields but this should give you a basic working knowledge. Many of these areas will be covered in more depth throughout this book but you may wish to pick up other books and magazines.

Like many other fields the specific terminology does not remain the same. Terms become interchangeable, blurred and abbreviated. Slang becomes standard and standards change. Modern equipment replaces old technology and particularly in home video and microcomputers a single item may do the work of many things. When you mix two technologies that overlap a little and have borrowed from each other, things get even more confusing. Undoubtedly, I have missed a few things here or there but if you have read this far then you should have a good base of understanding. If nothing else, you can now talk about video intelligently at cocktail parties.

# Chapter 2

## Camcorders and VCRs

# Chapter 2
# Camcorders and VCRs

Not so long ago there was no home video equipment other than your basic TV. Video cameras and video recorders cost tens of thousands of dollars. Portable video equipment didn't mean you could toss it in the car and go on vacation, it just meant that it was movable (if you had a large van and a crew of five or six engineers). Now it seems like every other person has a VCR and you can rent movies in any place with a cash register. There has been an incredible growth in the variety of video equipment available to the home user over the past few years and it looks like the trend will continue.

This chapter is about camcorders and VCRs. If you are reading this you probably already own a VCR and perhaps a camcorder. Just to get a few things cleared up at the beginning, I'll talk about the various formats that are available and explain which are best for desktop video and why. There were other formats that came and went, but this list is only for formats that are still available today. In any field people have their favorites and this chapter is by no means exhaustive. Use your best judgment and shop around.

## Professional Video

**Quad decks**  The first widely accepted video decks were called Quad decks (quad referring to Quadruplex). These are the two inch reel-to-reel monsters still used in broadcast studios today. They are about the size of a large refrigerator with a large cost to match. Unless you are a millionaire you probably won't be interested in a quad deck. If someone offered to give me a used one I probably wouldn't take it because they need about $100,000 of external equipment to operate and the tape alone would cost you more than a brand new VCR.

**One inch decks**

The next video decks that came along were the 1 inch reel-to-reel decks. These were the first alternatives to Quad but have almost disappeared over the past 15 years. When introduced they were considered professional level equipment but didn't give the quality that a good home unit of today can. You can still find them in the back rooms of studios and schools and you can sometimes get very good prices for them. They are large, heavy, require a lot of other high-end gear to operate and not portable at all, but if you want to set up a small studio they are nearly indestructible and produce a pretty good signal.

There are newer 1 inch (type C) decks available today for heavy-duty broadcast use but they are so far away from those first one inch decks that they should be considered brand new technology. They are expensive, produce a great image, and have every bell and whistle known to man. If you are setting up a broadcast studio and want the best then you probably want one inch.

**U-Matic decks**

The next breakthrough in video decks were the U-Matic decks. These were the first cassette decks that used 3/4" tape. Many TV stations still use U-Matic because the quality is very good and they don't cost as much as Quad decks. An average U-Matic VCR will cost about $2,000. While the tape is fairly expensive (about $50 for 20 minutes) you can sometimes find used U-Matic decks for sale. U-Matic is probably the lowest cost broadcast-quality equipment out there these days. Improvements have been made to the U-Matic machines and now you can buy a U-SP deck. They are still not considered 'consumer' decks but if you have the money they are pretty nice units.

**Digital systems**

Finally, on the high-tech professional level there is D2, a totally digital system that claims almost infinite generations without signal degradation. Some experts feel that D2 will become the next broadcasters standard.

On the professional camcorder side there are Betacam (not even remotely related to home Beta units), Betacam-SP, M-2, Hawkeye, Recam (type M), and Quartercam formats. These units use the same tape sizes as their consumer cousins but record the video in component (separate luminance/chrominance) rather than composite. They also run at about 6 times normal speed giving a maximum twenty-minute play.

## Consumer Video

**1/2" tape**

The first consumer level video recorders were 1/2" reel-to-reel black and white units. Their quality and reliability were so bad that we are fortunate they died an early death. The next two advances in 'home level' video appeared at about the same time and they are VHS and Beta home VCRs. They both use 1/2" tape but not in the same size cassette. The two formats are incompatible in that you can't play a Beta cassette in a VHS machine and vice versa. However, the video signals that they produce are pretty much identical so you can copy from Beta to VHS or VHS to Beta. Despite manufacturers claims, Beta decks have only slightly better resolution than VHS. So slight that you would need an oscilloscope to see the difference.

In the beginning it looked like there might be a battle between VHS and Beta but VHS has pretty much won the consumer VCR war. Estimates are that over 90% of the VCRs out there today are VHS. I'll talk about these decks in more detail a bit later.

**8mm**

Not waiting for a decision in the consumer battle between VHS and Beta formats, companies came out with 8mm camcorders. These boasted lighter weight, smaller format and a higher picture quality than either VHS or Beta.

**Superbeta**

Trying to gain an edge, manufacturers of VHS and Beta equipment came out with improved versions. On the Beta side there is Superbeta which yields about a 25% improvement in picture quality (if a Superbeta tape is played on a Superbeta machine) and then ED-Beta (Extended Definition) which gives almost 110% sharper picture than normal Beta. On the VHS side there is VHS-HQ (High Quality) which, like Superbeta, must have an HQ tape played in an HQ machine to see the improvement. Then, to try and reduce the size of VHS camcorders they came out with VHS-C.

**Super VHS**

There is also S-VHS (Super VHS) and S-VHS-C which give about 75% sharper pictures but need a special TV or monitor.

**8mm HD**

Not to be out done the newest wave is 8mm HD (High Definition) sometimes called HighBand-8 or just Hi8.

## Picture Quality

In order of picture quality the formats are:

**VHS**     1/2" (regular VHS cassette) 240 lines of resolution. Lowest quality consumer level.

**Beta**     1/2" (smaller cassette than VHS) 240-250 lines of resolution. About the same video quality as normal VHS, perhaps a shade better.

**VHS-C**     1/2" (smaller cassette than normal VHS) 240 lines of resolution. A touch better than normal VHS just because they came out later.

**VHS-HQ**     1/2" (regular VHS cassette) 240 lines of resolution. Better video quality than VHS but still a long way from broadcast quality.

**8mm**     1/4" (smallest cassette about the same as an audio cassette) 260-280 lines of resolution. Because of the metal tape 8mm gives a better picture than VHS.

**Superbeta**     1/2" (regular beta cassette) 300 lines of resolution. 25% better video than regular Beta or VHS.

**S-VHS-C**     1/2" (same size cassette as VHS-C but not identical) 400-420 lines of resolution. Much better video signal than VHS and getting close to broadcast quality.

**S-VHS**     1/2" (same size cassette as VHS but not identical) 400-420 lines of resolution. Slightly better quality than S-VHS-C just because of head size.

**Hi8**     1/4" (same cassette as 8mm) 410-430 lines of resolution. Slightly better than S-VHS because of metal tape.

**ED beta**     1/2" (same size cassette as Beta but special tape) 500 lines of resolution. Excellent quality video.

By the time you read this there will probably be four or five newer, better, higher resolution video decks and camcorders out there.

While the professional models are getting less expensive and smaller the consumer models are getting more sophisticated. If you are one of those people who just play movies, then the features you will want in a VCR are slightly different than those you should look for if you plan to do desktop video. A few manufacturers have realized that there are quite a few people out there who want to do more than just rent and play movies. Since VCRs and Camcorders can be of the same format

I'll talk about both simultaneously. Assuming that you are interested in Desktop video here are a few pros and cons about each of the consumer formats.

## VHS

VHS is the most popular format for VCRs. There are a number of advantages and disadvantages to getting this format. Since it is the most common format it is a fairly safe bet that if you end up with a VHS tape almost anybody can play it back. It doesn't make much difference how good your production is if only you can play it back (unless that is all you really want to do). Another advantage to VHS VCRs is that you can use them to play back rented movies. If you get a VHS Camcorder you can use the camcorder as a playback unit or just pop the tape into your VHS VCR. If you have a complete VHS editing setup (two VHS decks or a VHS Camcorder and VHS Deck) you can eliminate a generation in the editing process and anything you can do to eliminate extra generations is a big advantage.

Up until recently you couldn't buy a VHS deck or Camcorder with flying erase heads (about the only way you can do 'glitch' free edits). There are now a few VCRs and a handful of Camcorders that have flying erase heads but they are usually a little more expensive. If you are thinking about buying VHS for a desktop video application you should spend the extra money for a unit with flying erase heads.

The disadvantages of VHS is that it has the lowest quality. VHS is the result of a lot of compromises. In order to produce an affordable home unit manufacturers cut as many corners as they could. A VHS machine produces a signal just barely good enough to give you a fair picture on a home television but not by much. VHS also has severe timing problems that won't normally show up on your TV but will show up when you get into desktop video.

Since a VHS deck 'slips' quite a bit that means no frame accurate editing. It also means that it isn't a very good idea to use the sync signal from a VHS machine as your reference (if the VHS deck is the slave and all your other equipment is genlocked to it then your whole production will have the same sloppy sync). It is also rare to find a VHS VCR or Camcorder that will accept external sync and only a few higher priced VCRs and editing VCRs will accept input from an external editor controller for editing purposes. There is a way around some of these problems by modifying your existing equipment. There are companies out there that will sell you controller hardware that you can install yourself or have them do it.

Another problem relating to this slipping is that currently there isn't a TBC (time base corrector) made that can correct a VHS signal, the timing is just too far off. This is the main reason why you can't do infinite dubs (copies) of cassettes in VHS format (or Beta or 8mm).

Each play back has an error rate of about 10%. So the practical limit is about four generations and even that is stretching it a bit.

On top of all this, it is possible that some of the more sophisticated VHS decks with HQ circuitry (a system that gives you a slightly better picture) can introduce other problems. If you have two decks with HQ they can cancel each other out by trying to enhance the same picture twice.

As a point of interest, this slipping is how the copy protection schemes work. Movie houses introduce electronic fluctuations to a tape which are just under the threshold of a normal television set. The TV can usually compensate if you are playing back the original tape but when you make a copy of the tape on another VCR the second unit copies the artificial fluctuations (thinking that they are just part of the signal) and adds its own. The TV compensates for the second VCR but accurately reproduces copied fluctuations. With double the fluctuations the TV can't keep up any more and your picture will get brighter and darker, brighter and darker.

With all these problems as part of the VHS format you can add one more when you start talking about VHS Camcorders: The size. VHS Camcorders are usually about twice the size and weight of other format camcorders.

**VHS-C**

As a subset of the VHS format there is VHS-C. The VHS-C idea was an answer to the Camcorder size problem. The VHS-C cassettes are smaller than a normal VHS cassette but use the same format. There are special cases available that will let you playback a VHS-C cassette in a normal VHS deck. Claims are that the VHS-C format is slightly better resolution but the difference is negligible. While the VHS-C format may have solved the size problem they are limited to recording only 20 to 30 minutes on a tape in SP (Standard Play) mode. Plus VHS-C Camcorders have been known to produce something called vertical distortion. Because the heads are so close together they can cause tape vibrations. Manufacturers are working on ways to correct this.

So does this mean you should throw away your VHS equipment? Not at all. There are ways around many of these problems. You are going to want two VCRs if you want to do any serious editing and you can use a VHS deck as a slave unit. You will also want to eventually end up with a VHS format tape in most cases and if you plan your production carefully the final VHS copy might only be third generation (good enough for home use but don't take it to a studio).

**Beta**

While Beta format enthusiasts will tell you that they get a better picture the differences are minimal. The biggest advantage to Beta is the size. Beta Camcorders are smaller than VHS Camcorders, but since the

Beta format has become less and less popular you might have a hard time finding equipment. If you have Sony beta equipment, you may be able to connect everything together since that has always been one of Sony's selling points.

Beta has most of the advantages of VHS and also most of the problems. If you are thinking about purchasing new or used Beta equipment it might be better to look at some of the other formats first.

**8mm**

The 8mm format was developed early, particularly for Camcorders. 8mm offers a better quality picture than VHS or Beta. The Camcorders are about half the size and weight of a VHS unit. But perhaps the most appealing feature of 8mm equipment is the flying erase head. These days most 8mm equipment comes with flying erase heads which are almost essential in desktop video (you should still check the specifications before you buy).

The size advantage of 8mm camcorders is worth serious thought when deciding on a unit. If you are going to be dragging a camera around with you on shoots would you rather carry 2 pounds or 5 pounds? Will you be more inclined to grab a small hand-held unit or an on-the-shoulder unit? How long can you hold a large unit steady? While these are not really 'technical' considerations they are part of the overall desktop video picture.

There are disadvantages to the 8mm format though. A VHS Camcorder cassette can be played on a VHS VCR eliminating a generation while you may have to dub from an 8mm Camcorder to another deck. It is possible to use the Camcorder as a slave in the editing process but many 8mm Camcorders don't have external controls or the controls are limited.

Another disadvantage to 8mm Camcorders is that in order to squeeze everything down to fit on 8mm tape, manufacturers did away with a separate audio track. Instead, they record the audio information in with the video. This makes it almost impossible to perform an audio dub (see below) without adding a generation.

If all you have is 8mm equipment you won't be able to share tapes with very many people and you can't rent movies in 8mm format either.

**Superbeta**

The next step up would be Superbeta. Superbeta gives you a better picture, and it is smaller and lighter than VHS. When you get around to dubbing down to another format, your overall quality will be better than going from 8mm to VHS or VHS to VHS. There are a few very nice Superbeta decks out there with all the extras (if you want to pay the price) but the selection is limited. If you can find the right connectors you can use the Superbeta deck as a slave unit in an editing suite or use two Superbeta decks.

While you will get a better picture with Superbeta the availability of peripherals may be a problem. The odds are that you will also have to dub down to another format because there aren't that many people out there with Superbeta decks. Finally, like regular beta, I have doubts about the life expectancy of Superbeta equipment.

**S-VHS and S-VHS-C**

The newest wave of high-end consumer (or low-end professional) equipment is S-VHS (Super VHS). S-VHS features about twice the picture quality of VHS. The decks and Camcorders are usually loaded with professional features like flying erase heads. The sync and video quality is much better than normal VHS which means there is less degradation with each generation. S-VHS equipment can also record and play in VHS format. S-VHS equipment has more professional accessories available like editor controllers and switcher-faders. S-VHS is about the top of the line in VHS equipment.

But there are problems with S-VHS too. While the stability of the signal is much, much better than normal VHS is still just barely good enough for professional broadcasters. There are still problems with frame accurate editing. It can be done but it will cost you.

S-VHS can produce a very good image, much better than your TV is capable of reproducing, which means you need a special monitor to see the 'Super' part. And while S-VHS may become a standard some day (there are newer TV sets on the market now with a switchable S mode), there are not very many people out there with S-VHS equipment today.

The biggest disadvantage to S-VHS is the cost. Everything from the Camcorders to the VCRs and special TVs will cost about twice as much as normal VHS equipment (the least expensive S-VHS camcorders start at about $1600, VCRs will run you about $1200).

The S-VHS-C Camcorders have the same advantages and disadvantages of S-VHS and VHS-C. Better quality, smaller size, limited recording time per tape (although they are working on this), price, compatibility, etc.

One more note on the S-VHS format. While the claims are for 400 lines of resolution you should know that they use a different formula to calculate those 400 lines.

**Hi8**

The other high-end consumer/low-end professional format is Hi8 (High Band 8mm). Like the S-VHS equipment picture quality is very good (actually Hi8 has a slightly better picture than S-VHS). Hi8 also has limited degradation of signal from generation to generation, professional quality features, is usually switchable from Hi8 to normal 8mm formats, and requires a special TV to appreciate the 'Hi' part.

Hi8 Camcorders are also much smaller and lighter than VHS or S-VHS. Hi8 (and S-VHS) decks and camcorders get some of their better picture quality from the fact that they record and output the video signal in Y/C (another term for non-composite, luminance/chrominance). This means that you can output your Hi8 signal to an S-VHS deck and vice versa. Hi8 also makes use of the fact that it uses a metal base tape rather than an oxide base.

On the down side Hi8 is HiPriced. The least expensive Hi8 Camcorders and VCRs will cost you about $2,000 on up. Also I wouldn't expect to see a lot of rental movies in the Hi8 format in the near future. And you will probably have to dub down eventually.

**ED Beta**

Finally, at the very top of the "consumer" line is ED Beta. If you can afford it definitely go for ED Beta. The picture quality is very high, there are lots of professional level peripherals, you can produce near-broadcast quality. Camcorders are small and lightweight, and ED Beta will stand up to more generations than other formats.

The biggest disadvantage (apart from the fact that few people own ED Beta equipment) is the price. ED Beta is close enough to broadcast, professional level equipment that the prices are at broadcast, professional levels too (ED Beta decks cost over $3000, camcorders are over $7500) You also won't find dozens of models to pick from or too many peripherals in your local K-Mart.

So what does all this mean to an aspiring Desktop Video producer? Which systems are the best for the money? What do you really need and what can you do without?

Like everything in this world you get what you pay for. If money is no problem then you should be looking at the ED-Beta, S-VHS or Hi8 equipment. This equipment will give you a better signal, higher quality and more professional features. You can always do a final dub in VHS format if you need to (and you probably will, unless you are only going to be playing back on your own equipment).

Of these three, I would recommend the Hi8 format for a few reasons. The size of the Camcorders is always a consideration and Hi8 units are smaller and lighter. Next, the quality of the output is slightly better than S-VHS. While you could dub S-VHS to Hi8 the quality will always be slightly less than the original which means a Hi8 to S-VHS dub will be close to 100% S-VHS quality but an S-VHS to Hi8 dub will definitely be less than 100% Hi8 quality. While Hi8 doesn't have the resolution of ED-Beta it is about half the cost plus there are and will be more brands and peripherals to pick from. Another reason I would favor the Hi8 is that companies are talking about after-the-fact time coding units and in-camera time coding which should give frame accurate editing capabilities. And metal tape has yet to be fully exploited while plastic based tapes have been pushed about as far as they can go.

S-VHS would be the next choice because of the quality and the wide variety of manufacturers and peripherals.

Next, the choice would have to be regular 8mm Camcorders. They are small, light and give a better picture than a VHS Camcorder. Most of them have flying erase heads which make for clean in-camera editing. I would favor 8mm over Superbeta just because of the number and variety of companies producing 8mm equipment. Superbeta will give a better picture but is expensive and you won't have the selection that other formats offer.

Next, would be the VHS Camcorder-VHS VCR combination. You can find VHS machines with flying erase heads, you can shoot much longer than with VHS-C Camcorders, most people out there will be able to play back a VHS tape, and you can usually avoid an extra generation during the editing process.

At the bottom of my list would be VHS-C and Beta. The VHS-C time constraints and vertical distortion problems outweigh the size advantages. Beta, in my opinion, is not long for this world and I am not a big believer in buying into a dying technology.

But there is another possibility. The Mix and Match philosophy. Take the best of each and make due with what you already have. If you can afford an extra machine for the final dub and can connect your equipment in a usable way then it doesn't matter which components you have in your system. You could always do your work in one format and pay someone else to do a final dub for you or if you are only doing things for yourself then format isn't going to be a problem. The thing about video is that in order to put a picture on a normal TV the signal has to meet certain specifications. That means you could use a S-VHS-C camcorder to shoot with, dub over to Hi8, use the Hi8 as a slave and an ED-Beta deck as the master and then dub down to VHS for your distribution copies. Whatever system you decide on, there are a few minimum requirements for desktop video applications.

| **Flying erase heads** | The very first priority is to get a master unit with flying erase heads. This can be a VCR or Camcorder usable as a VCR. If you plan to use the camcorder as a master then it has to be able to record video from an external source, not just through the lens. Not all camcorders will do this.

There are a few camcorders and VCRs out there with something called a "Zero Frame Editing" feature instead of a flying erase head. This system works fine if all you want to do is assemble edits but it cannot do insert edits. While it is possible to put together a production using only assemble edits, the more flexibility you have the better.

Why are flying erase heads so important? The simplest explanation is that you can't do clean, glitch free editing without them. Here is a quick semi-technical explanation of why.

In an audio tape recorder the magnetic heads that record, erase and playback are fixed while the tape travels past them. Video recording has a lot more electronic information to store on the tape so while you could use stationary heads too, you would need lots and lots of tape. One solution used by most VCRs is to have the record and playback heads in a spinning drum mounted on a slight angle. As the tape travels around the drum each head lays down (or plays back) a "track" diagonally on the tape. This way more information can be stored on less tape. If you could see the magnetic tracks on a video tape (which you can't) you would see thousands of diagonal lines. Each line representing one field of a video image (remember a video picture is one frame made up of two fields). See Figure 1.

Lets say you have a piece of video tape with a scene recorded on it already. If you numbered every field and also numbered each millimeter of the tape, you would find that the track for field number 1 starts at millimeter 1 but might end at millimeter 11. The track for frame 2 might start at millimeter 2 and end at millimeter 12, etc. If you cut straight across the tape at millimeter 12 you would cut across many field tracks. The field one track would be complete, but the field two track would be cut off at the end with only 9/10ths finished, field three 8/10ths, etc. If you tried to play this back, the first field would be fine but the second field would end before the TV finished drawing the picture. This is more or less how a stationary erase head works.

If you tried to cut from one scene to another at millimeter 12 with a stationary erase head you run into problems. At millimeter 13 you start field track 1A, at 14 you start 2A, at 15 3A, etc. But you haven't given fields 2 through 10 a chance to finish yet. When the tape is played back field 1 is fine, but fields 2 through 10 are cut short. When the monitor can't figure out where it is it goes a little crazy, this is the 'Glitch'. In this case there would be confusion for a short time because the monitor |

won't settle down until it has seen at least two clean fields (probably a little more because it takes a second or two for the VCR and monitor to settle down once it finds the new signal). See Figure 2.

A flying erase head is mounted on the same drum that the record/playback heads are on. Rather than erasing across the tape it erases diagonally. This means that each field track has a chance to finish. See Figure 3.

You might think that like audio tape you could just record over the old information without any kind of erase head at all. You can, to a degree, and there are some VCRs that try to do just that. The problem is that it is almost impossible to erase everything this way.

Beyond flying erase heads there are other features that you might want to look for in a camcorder or VCR. While they are not absolute necessities they can make life a great deal easier for the desktop video producer.

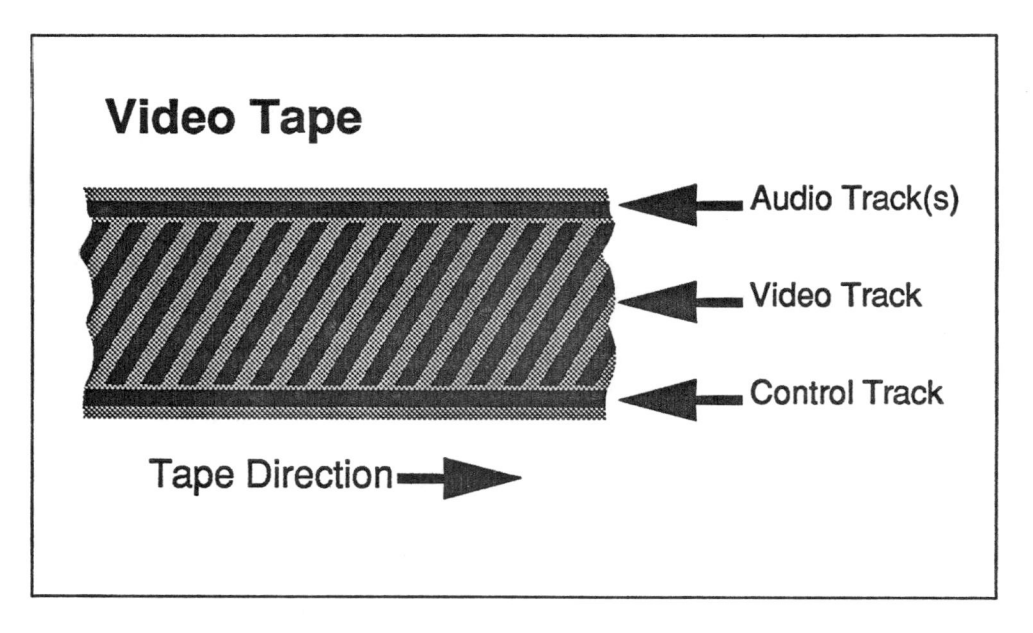

*Figure 1*     *Video Tape*

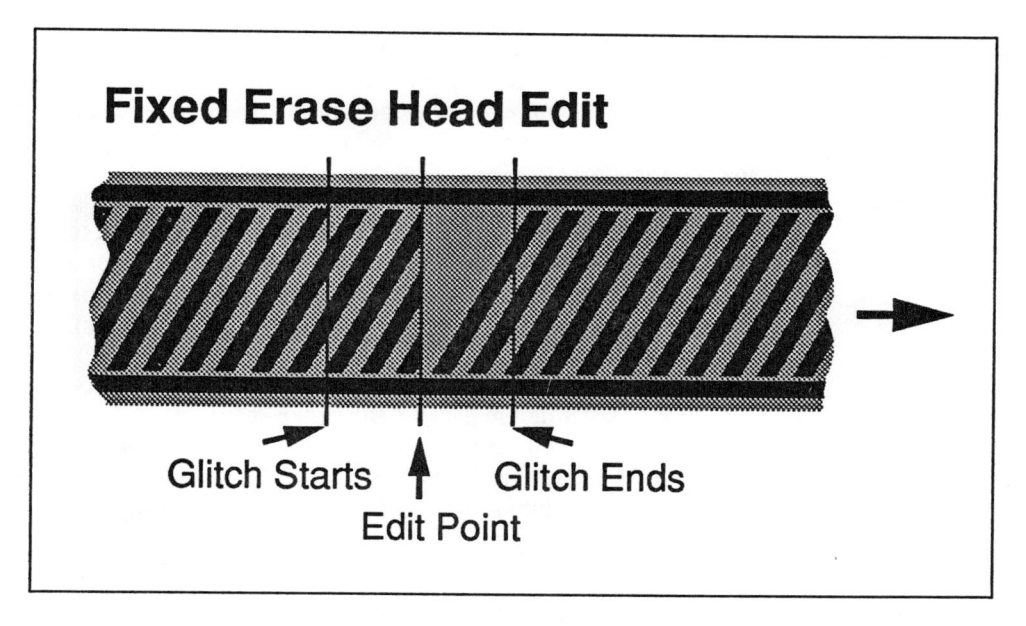

*Figure 2     Fixed Erase Head Edit*

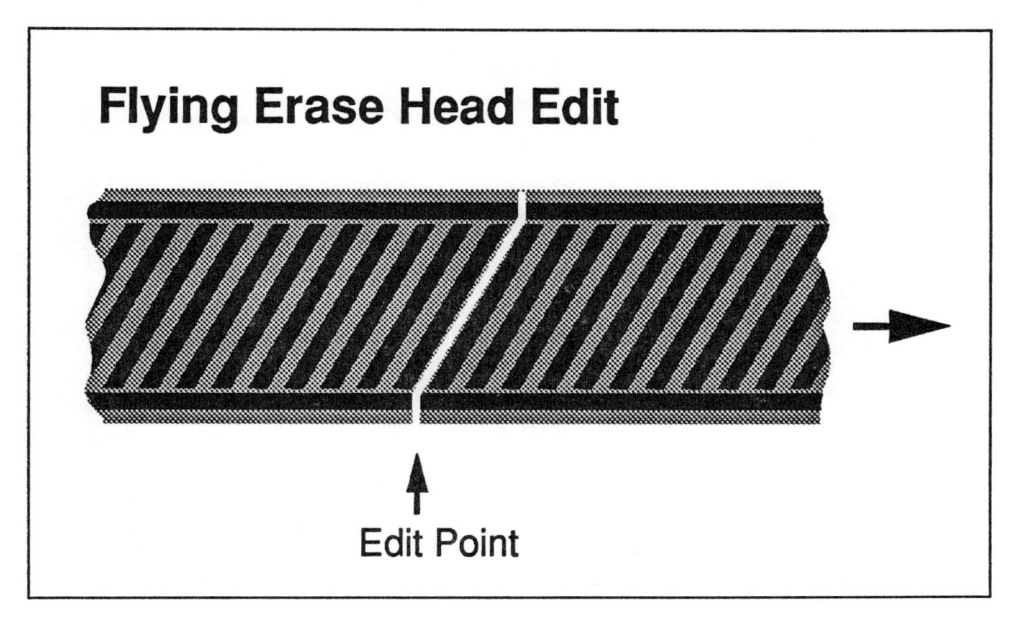

*Figure 3     Flying Erase Head Edit*

**External control**

Just because a VCR or camcorder has a remote control unit does not mean that it can be used for clean editing. Many camcorders these days come with external control L, control S, or another type of external control jack (input) that will allow the unit to control or be controlled by another VCR. To see where this ability comes into play lets look at a typical assemble edit.

Lets say you have scene one recorded on a VCR and you wish to add scene two which is recorded on the camcorder. The Camcorder is the slave unit and the VCR is the master unit. These are the steps you would use to perform the edit manually (without external controls).

1. Find the end of scene one on the master unit.
2. Put the master unit in pause at the end of scene one.
3. Put the master unit in record/pause. (The master unit is in record mode but the tape is not moving.) Note: Some older VCRs will not let you go into record mode unless the unit is stopped. The odds are that these units won't have flying erase heads anyway but you can still put these units in pause, press stop, press record, then play again. Hopefully the unit will still be in pause.
4. Find the start of scene two on the slave unit. (The scene you wish to add.)
5. Put the slave unit in pause at the beginning of scene two.
6. Simultaneously take both units out of pause. (By pressing the pause buttons with your two hands.)

With an external control cable, Step 6 is performed automatically. On some VCRs when you take the master out of record/pause, it also takes the slave unit out of pause. The thing to remember about external control jacks and cables is that the units must be compatible. It is usually best to stick to equipment from a single manufacturer. Even then there is no guarantee that two units will work together. Check before you buy.

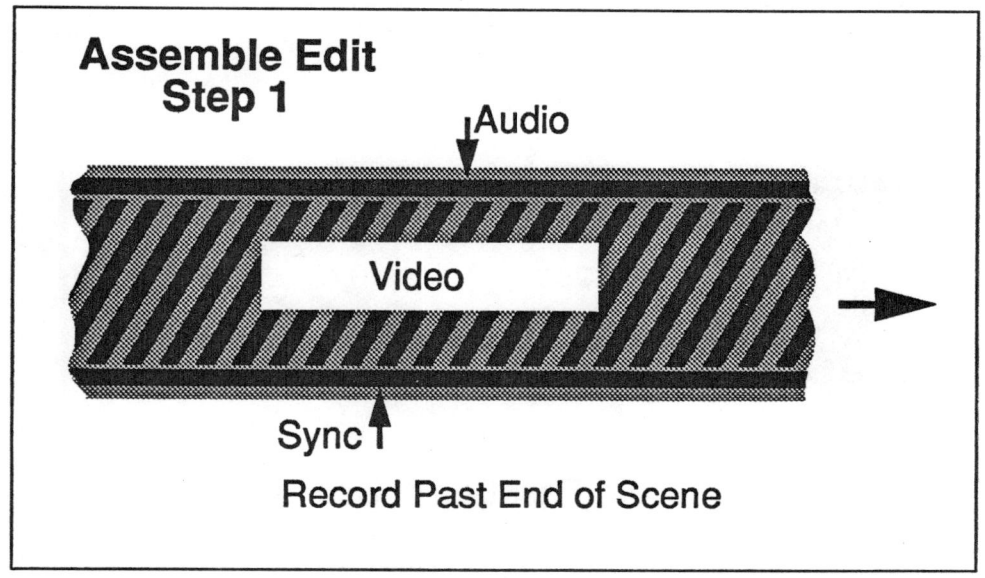

*Figure 4*     *Assemble Edit Step 1*

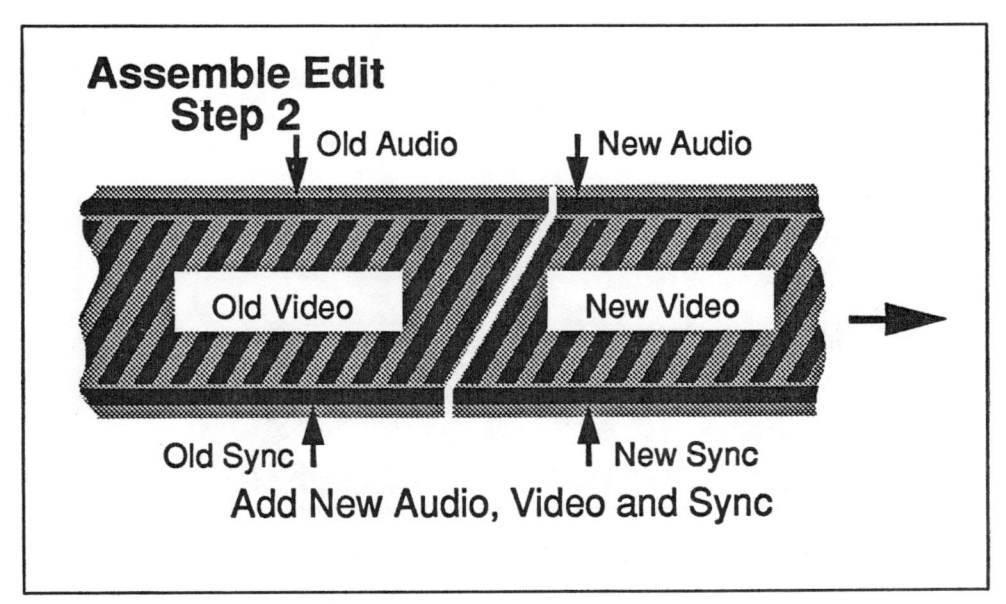

*Figure 5*     *Assemble Edit Step 2*

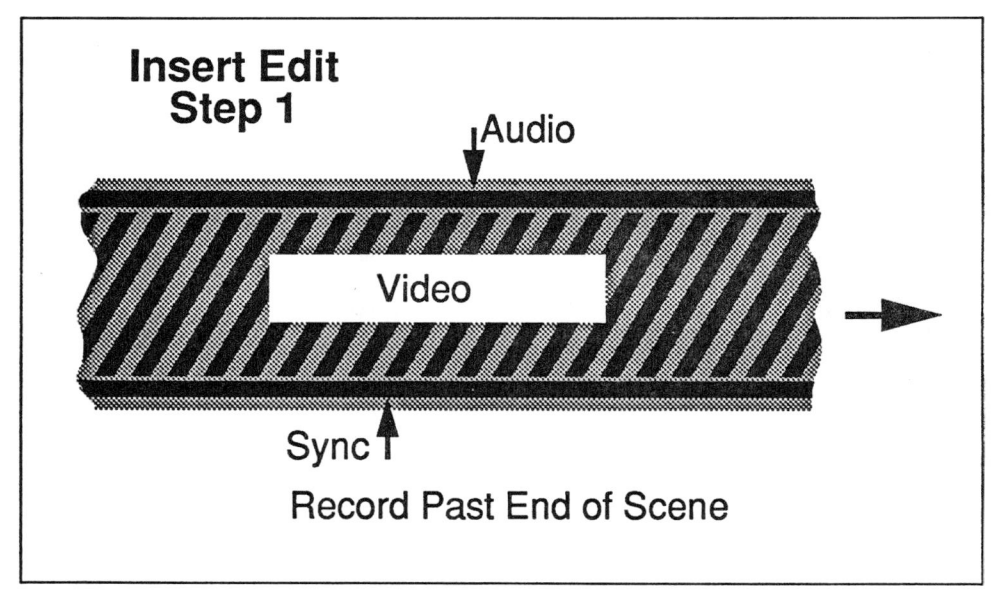

*Figure 6      Insert Edit Step 1*

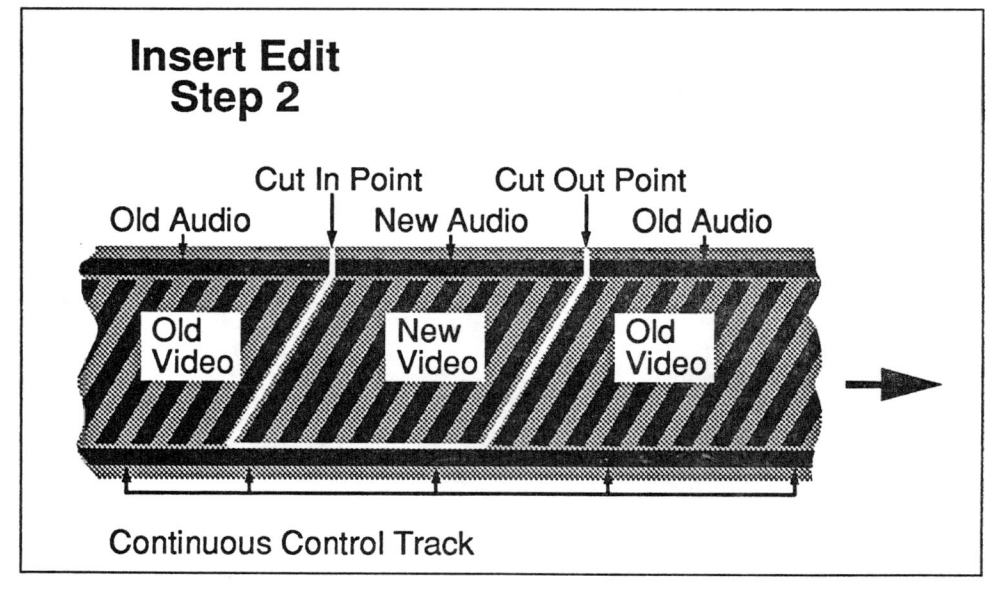

*Figure 7      Insert Edit Step 2*

External control can also let you control both the master and slave unit with an editor/controller. Some editor/controllers will do much more than just take the units out of pause for you. They let you control just about every function of the master and slave units from a separate controller. Some can also let you perform pre-roll edits. The advantage of a pre-roll edit is that both units are already up to speed when the edit is performed, giving you a much cleaner edit. Here are the steps using an editor/controller performing the same assemble edit with pre-roll.

1. Find the end of scene one on the master unit and press a button on the editor/controller. The editor/controller stops the tape, rewinds it a precise number of seconds (anywhere from 2 to 10 seconds depending on the VCR), then puts the master unit in pause.

2. Find the beginning of scene two on the slave unit and press another button. The editor/controller stops the tape, rewinds it the same number of seconds and puts the slave unit in pause.

3. Press another button (usually called "perform edit", "execute", or something like that). The editor/control starts both decks simultaneously and at exactly the right point throws the master unit into record without going through pause.

The advantage of using an editor/controller should be obvious. It does many things for you and does them more precisely than you could ever do by hand. In professional studios the editor/controller can perform frame accurate edits and it can even make sure that the edit happens between frames so that even the slightest glitch happens where you can't see it. What do you do if you already have a VCR and camcorder but they don't have external control capabilities or they are not compatible? There are three solutions. First, you could go out and buy different equipment. Second, while it isn't as accurate you can still do edits manually. Third, there are companies that can modify almost any VCR so that it will accept external control, even let you use the Amiga as an editor/controller. See Appendix A.

**Audio Dub**  Audio Dub is another feature that you should seriously consider when shopping around for equipment. Simply stated, audio dub lets you go back and change the audio track on a tape without effecting the video. This feature can save you an extra generation (and anything that saves extra generations is worth while). Without audio dub you have three options. First, just make due with the sound on the tape as it is. Second, only modify the sound on individual scenes during the editing process. Third, add an extra generation just for the audio.

**Special Extras**  Many camcorders these days are offering a wide variety of extras and special effects like time lapse, built-in character generators, super-fast shutter speeds, 12-1 zoom lenses, and on and on. While some of these features are nice to have most are not critical to desktop video. A few words and thoughts on these features.

**Flying erase heads**  Flying erase heads in a camcorder are also nice to have for a few reasons. If you are planning to use the camcorder as part of your editing studio, you may wish to use the unit as the master and the master unit must have flying erase heads. Another reason for having flying erase heads in the camcorder is that with careful planning you can do much of your editing during the shoot. Camcorders without flying erase heads suffer from glitch problems between scenes, which means you must let the unit get up to speed before any action takes place. Back in the editing suite there is no way to eliminate these glitches, even with a master unit that has flying erase heads. You must wait until the recorded signal stabilizes or you just end up re-recording the glitch.

**Character generators**  Built-in character generators are so-so in quality, take time to set up and the computer will do a much better job in the end.

**Zoom lenses**  12-1 zoom lenses are nice but not absolutely necessary. However, you should look for a camcorder that features at least a 6-1 zoom. Macro-zoom lenses are also very nice. They allow you to focus on something as close as a few inches from the lens. A macro lens can substitute for a frame grabber (in a crude way) by pointing the camera at a photograph. Mixing still photography with action video can produce some very nice effects. Adjustable zoom speed is something you will definitely appreciate but can live without. With a steady hand you can zoom manually at any speed you like.

**Low Lux**  Low Lux numbers are also nice but keep in mind that just because it claims to be able to shoot with the light of one candle doesn't mean the quality will be any good. Get used to the idea of using extra lighting.

**Back lighting**  Back lighting mode is an extra that can be accomplished simply by manually adjusting the white level or F-stop settings (providing the camcorder will let you).

**Fade Control**  Fade controls (usually in and out, black or white) are also very nice extras to have in a camcorder but can usually be simulated with a genlock/computer combination.

**Time Lapse**  Time lapse mode is neat but let's face it, how often do you need shots of flowers opening in slow motion?

| | |
|---|---|
| **Animation mode** | Animation mode (a variation on time lapse mode) is not only nice for doing claymation (do it yourself Gumby and Pokey) but could also be very useful for doing single cell animation with the computer. It should be noted that the camcorder isn't recording just one frame at a time. It records for a second or two, stops, rewinds a little and puts itself in record/pause mode ready for the next shot. Since home equipment isn't capable of single-frame accuracy, if you read the fine print you will see that most units only claim from 4 to 10 frames accuracy and their animation modes are as high as one or two seconds (30-60 frames!) a shot. This may mean a little jerkiness and it will also mean experimenting a bit to get animations happening at the right speed. If you are thinking about single-frame animations on the computer, make sure that the camcorder you choose has manual control of the animation mode (some just go off every twenty seconds or so which may not give you enough time to set up each shot). |
| **Shutter speed** | Superfast shutter speeds are items offered on many camcorders these days. They will let you freeze the motion of a golf club in mid swing or a diver in mid air but they require a great deal of light, need a playback unit that will freeze that action, can give jerky playback at normal speed and have limited use. If you plan to do a lot of action tapes then this feature would be a must but most of the time you will never need it. |
| **Color viewfinders** | Color viewfinders are pointless. In fact, a black and white view finder will give you a sharper image and it is easier to focus with. |
| **Special Effects** | Negative/Positive special effects you might use once and say "that was neat," then you will then never use it again. |
| | Undoubtedly there will be dozens of new features offered in the near future. Special effects in the camcorder are nice but many of them can be done better with the computer. In general, any extra feature that enhances the video quality or that will save a generation is valuable. As far as VCRs go, again there are dozens of extras you can get if you want to pay for them. |
| **Flying erase heads** | Flying erase heads are a must if you plan to use the VCR as a master unit. If you haven't figured out by now that flying erase heads are important go back and read this chapter again. |
| **External control jacks** | External control jacks are very valuable, if they match your other equipment. |
| **Multiple heads** | Multiple heads mean better still and slow motion playback but won't increase the video quality beyond the format limitations. |

| | |
|---|---|
| HQ circuitry | HQ circuitry will enhance the video signal to a degree but don't expect miracles. |
| Jog/Shuttle Controls | You will usually only find jog/shuttle controls on the more expensive editing VCRs. These let you control the tape movement forwards or backwards at variable speeds by hand. This can save you a lot of time during editing. Most VCRS will let you freeze-frame (pause) and then step advance one frame at a time, but they don't backup one frame at a time in pause. Of course, this is more of an annoyance than a problem. |
| On screen programming | On screen programming is nice when you are using your VCR for off the air recording but doesn't really have any meaning for desktop video applications. |
| Audio dub | Audio dub is a VERY nice feature to have because it eliminates a generation during the editing process and anything that eliminates a generation is worth while. |
| Editing | Editing features can be simple to extravagant. There are editing VCRs available for home use that will cost you about twice what a normal VCR will but if you can afford one they are worth it. |
| Stereo | Stereo features are nice but not necessary. Eventually most VCRs and TVs will be stereo but until then it isn't critical. It might be better to trade stereo capabilities for flying erase heads or other features more critical. The only time when stereo might be desirable is if you can use one channel for audio dubbing. |
| Tripods | Beyond the features found in the camcorder or VCR there are extras that you should consider purchasing if you don't already have them. A good tripod will make a world of difference for your productions and your shoulder. It is worth it to spend the extra money for a good tripod. You can use a film camera tripod but remember that most still cameras weigh only a fraction of what a camcorder does. Get a good sturdy tripod that has smooth panning action. A film camera doesn't need to pan smoothly but a camcorder does. |
| Lighting | Extra lights are worth buying. You don't have to get special video lights but if you want to mount the light on the camcorder (not a great idea) you need one that will fit. If possible mount your lights off to one side and above the camera. This will avoid that "flat" look you get with a camera mounted light. In a studio there are special lights for everything. In order of importance the extra lights you might consider getting are: |

Key light   This is the main light shining on your subject. Usually mounted above and to one side of the camera. Ideally the key light should be mounted 30 to 40 degrees above the camera and 20 to 45 degrees to one side.

Fill light   This light is also for your main subject. Mounted the same way as a key light only on the opposite side of the camera. A fill light is more diffuse than a key light. In a pinch use the light coming through a window or take the shade off a regular room light as a fill.

Back light

Again this light is aimed at the main subject only it is mounted directly behind and above the subject from 45 to 75 degrees up from horizontal.

Set light   This light, as the name implies, is for illuminating the set or background.

Modeling or Kicker light

Like the back light the modeling light is for illuminating the main subject from behind only. It should be mounted high and off to one side. The only hard and fast rule about lighting is - the better the lighting the better the video quality.

**Filters**   Filters for your camcorder perform two functions. First, they can improve, distort or modify the image and second, they protect your main lens from accidental damage.

**Microphones**   Microphones come in all shapes and sizes for all kinds of situations. Usually the microphone that comes with your camcorder is just fair. You might wish to look into getting additional microphones depending on your application. If you will be doing studio work then lavalier and boom mikes are important. For doing voice overs you might consider a table stand directional or unidirectional mike. For doing interviews in the field there are numerous hand-held mikes and even telephoto mikes (basically a shotgun mike). For long shots you might think about a radio microphone.

**Audio mixer**   An audio mixer is almost a requirement if you plan to do any kind of extra sound, music, voice overs or sound effects. While a production can be done without these extras you would be amazed at the difference a little background music can make.

**Batteries**   Extra batteries for your camcorder, lights, etc. There is nothing worse than having your battery go dead in the middle of a shoot. You should also bring your battery charger with you just in case. It is better to take an hour time-out than have to come back the next day.

**Tape**  Extra tape is essential. As far as which tape to buy obviously you should stick to name brand tape designed for that format (don't use VHS tape in an S-VHS unit). But as far as tape quality goes there isn't much difference from the top of the line to the bargain basement until you get to the 50th play. Of course, when working in VHS, Beta or 8mm even a slight improvement may be worth while so you might want to stick to a manufacturers top quality HG, EXG, Pro, or whatever they call it. Scotch (EXG Pro, EXG Hi-Fi, EXG Camera and EXG), TDK (HD-X Pro, E-HG and HD), Fuji (Super HG and HG Hi-Fi), Maxell (RX and HGXHF), and Sony (Pro-X, ES-HG and ESX-HiFi) are all good tapes. Radio Shack, Certron and J.C. Penney aren't terrible but should probably be avoided. Ironically, the tape you use for desktop video will probably only get played a few times during recording and editing while the tape you use to record the Saturday morning cartoons for the kids will get played to death. So buy the top quality tape for the kids and whatever HG tape you are happy with for the serious stuff. By the way, video tape is usually higher quality than audio tape to begin with. So if you see special "high-fidelity" or "stereo VCR" tape, the manufacturer is just trying to get a few extra bucks from peoples ignorance.

**Tape tips**  Store tapes on their edges not flat. Avoid heat, moisture, changes in temperature. Fast forward and rewind tapes at least once a year (it is also a good idea to do this just before using a tape for a master). VHS, and Beta tapes perform better after about 10 plays (the heads polish the tape a bit) while other format tapes work best straight out of the box. Label every tape immediately after you use it (Murphy's video tape law states "when you accidentally record on an unmarked tape it will be the most valuable tape in your collection, while 'junk' tapes last forever").

**Carrying Case**  Finally a good carrying case for all your gear. Expect to pack a lot of extra stuff not just the camcorder.

**Stop Watch**  You will eventually need to pick up a stop watch for editing. Since you are only timing one thing at a time a cheap one is all that you really need. Home editing is not precise down to 1000ths of a second, but you will need to time scenes and finding edit points is much easier with a stop watch.

## Summary

Computer people and video people have one thing in common. They never have everything that they want. If you are on a strict budget be sure to allow a few hundred dollars extra for cables, tape, diskettes and all the other little things. Any level, beginner, intermediate or advanced desktop video producer can benefit from a few books and magazines (a list of publications appear at the end of the book). Shop around before you buy anything and if you plan to buy equipment through the mail, be careful. Many things covered in this chapter will be covered in more detail later so you should read through the entire book before you rush out and plunk down thousands of dollars. Hopefully, we have covered a lot of the video equipment.

# Chapter 3

## Genlocks

# Chapter 3
# Genlocks

So now you have all your video equipment. The camcorder, the VCRs, the lights, tape, tripod and a thousand other items. You want to start doing all those special things that desktop video promises. You want to put titles on your videos, special effects, music, computer animations and graphics. What is the next step?

The next step is the computer. Since this is a book about the Amiga and desktop video the choice of computers is easy. Why is this a book about desktop video and the Amiga? Because the Amiga is ideally suited for video work. When you combine the graphic capabilities, the animation, the sound and the peripherals that are available, the Amiga stands out as a perfect choice.

If you don't already have an Amiga, there are a few things to know before you rush out and buy one. And if you already own one, there are a few things that you should know before you start trying to plug your Amiga into your VCR. Yes, this is a chapter about Genlocks and I will get to them shortly but first a bit about the computer itself. If you already know everything there is to know about the Amiga then you can skip ahead.

## The Three Amigas

At the moment there are three basic Amiga models. The A500, A1000 and the A2000. There are two other models called the A2000HD and the A2500 but these are essentially A2000s with extras already built in. Right now the A500 can be purchased for under $1000 (including the monitor and extra memory module). The A2000 will cost you about $2000 with a monitor and extra memory. The A2000HD will cost you about $2500, and the A2500 about $3000.

**Amiga 1000**  The first Amiga was the A1000. Many of you already own one of these and even though they haven't been made for a few years you can sometimes find them for sale. There is little difference between the A1000 and the other models as far as the operating system goes.

Software that runs on one machine will usually run on all of them. The only software difference is that the A1000 required inserting an extra disk (the Kickstart diskette) when you booted the system (turned it on and got it ready to run). After the kickstart program loaded, you had to insert the Workbench diskette to finish booting the A1000. Once you got it up and running it behaved just like an Amiga 500 or 2000. The biggest hardware differences are in how you connect the peripherals. The A1000 uses a different printer cable, the expansion port is different, and you can't open up the A1000 to plug in boards like you can with the A2000. While there are a few genlocks available for the A1000 specifically, and many genlocks will work with any Amiga, most manufacturers are no longer making peripherals for the A1000.

**Amiga 500**  The A500 is the baby brother of the A1000. It was designed and built for the "home market" which means it is less expandable, stripped down (internally) and less expensive. When you boot up an A500, you must first insert the "Workbench" diskette (the programs that make the Amiga behave like an Amiga) Once the A500 is up and running, it works just like the other Amigas as far as the software goes and almost like an A2000 as far as the external peripherals go. The biggest difference between the A500 and the A2000 is that you can't open it up and plug in boards. Since the A500 is the hottest selling Amiga, most manufacturers that make peripherals for the Amiga will be doing versions of their products for the A500. However, when it comes to some of the more expensive video peripherals, some manufacturers don't feel that an A500 owner will be willing to spend the extra money.

**Amiga 2000**  The Amiga 2000 (including the 2000HD and 2500) is software compatible with the other Amigas. Like the A500 you don't need a kickstart diskette to boot the system. Almost all external peripherals that work with the A500 (printers, modems, external genlocks, etc.) will work with the A2000. But the A2000 is designed to be opened and upgraded with internal boards. Most hardware manufacturers make an A2000 version of their products because they assume A2000 owners are more serious about computing, have more money to spend and it is easier to manufacture peripherals if you don't have to put them in a separate case (like you have to for an A500 peripheral).

To the computer novice the Amiga offers another advantage over other makes. It is easy to use. Like the Apple Macintosh the Amiga uses a graphic interface and mouse control. To run programs you just pop in a 3.5" diskette, when the program icon appears on the screen, point at it and click the mouse button. Obviously, there is more to it than that but it is easier to use than most computers.

Finally, the Amiga has advantages over other computers when you start talking about video. It can be expanded to 9 megabytes of RAM (an IBM can only be expanded to 640 kilobytes of RAM) which means the

Amiga can do more with graphics and animations. The Amiga can produce an overscanned picture which is important for video. The Amiga can be made to run faster than most computers by simply plugging in a different chip (the 68020 or 68030 with arithmetic co-processors). This is very important when you start doing very complicated graphics (like ray-tracing). The Amiga can be run in an interlaced mode like the signal for a standard TV. And, since the Amiga produces a 15.75 Khz analog RGB signal, it is much easier and cheaper to manufacture Amiga video devices than it is for the Macintosh or IBM.

As far as desktop video goes any one of the Amigas will work but the 2000 series has some advantages over the 500 and the 1000. Which ever Amiga you have or are thinking about buying, I would suggest buying a second disk drive and as much memory as you can afford. Memory expansion is crucial if you want to do animations or ray-traced graphics. One megabyte is about the absolute minimum but it is possible to do many things with less. The extra disk drive will save you a lot of disk swapping.

If you have the money, buy an A2500, A2000HD or A2000. If you don't have the money, buy an A500 and the extra memory. When you have it, read through the manual and play around with the computer until you are familiar with how it works. Now on to genlocks.

## Encoders, Sync Generators and Genlocks

When the Amiga was created the engineers must have had desktop video in mind. The video signals coming straight out of the Amiga are almost up to NTSC standards but not quite. The A1000 outputs a composite color video signal (sort of) while the A500 and A2000 output RGB analog non-composit signals (sort of). It is possible to just plug the Amiga straight into a VCR and record whatever appears on the screen (you will only get color if you record from an A1000). The trouble is that you won't get a very good picture. That is where the encoder comes in.

**Encoder**  In order to get a good, clean signal from the Amiga you need something called an encoder. An encoder is a small, fairly inexpensive device that takes the video output from an Amiga 500 or 2000 and produces a color composite signal. With a good encoder you could do a number of things with the Amiga. You could record graphics, animations, or just about anything that appears on the Amiga screen. An encoder would turn your Amiga into another video source, like a camera or VCR. This would be enough to do titles, illustrations, pictures and animations and record them directly onto video tape.

While encoders are primarily for using the Amiga with a composite monitor, there are a few desktop video uses. You could use an encoder for titling but you would have to do straight cuts. For example, you could start your production with a title screen from the Amiga, cut to another title screen, cut to video from a camera, cut back to graphics on the Amiga, and so on. With the right software you could fade from one Amiga image to another but you couldn't fade from the Amiga to an external video source. Another application for an encoder only setup would be for transferring Amiga animations to videotape.

There are a few encoders available for the Amiga. Unfortunately, while they will turn your Amiga 500 or 2000 signal into something close to true video they are mainly for using the Amiga with a normal TV or composite monitor, not for desktop video. For the most part the signals they produce are very poor. Without an encoder, about the only thing that you could use the Amiga for in desktop video is music and sound effects. (See Chapter 9.)

If all you really need (or can afford) is an encoder and nothing else, the only unit I can recommend at this time would be the CMI VI-500 or VI-2000 from Creative Microsystems, Inc. At $79.95 the units are priced right. I have not seen the results on a scope but have heard these units do a pretty good job. CMI manufactures external units for the 500, 1000 and 2000, an internal unit for the 2000 (that uses the video slot) and for $20 extra you can get units with RF out (for use with regular television sets).

But even with a good encoder you wouldn't be able to overlay your graphics onto other video or dissolve from the Amiga to another video source. If all you have is an encoder you couldn't have titles appear on top of other video for the same reason that you can't mix video from two VCRs or dissolve or wipe from one VCR to another.

I know, you are saying to yourself, "Wait a minute. I see video on top of video all the time. Why can't I do that too?" The answer is vague, you can and you can't. With home equipment you can cut smoothly from one video source to another (if you have a VCR with flying erase heads). You can also fade one scene to black then fade into another scene (again with the right equipment like a switcher/fader). But you can't be fading one scene in as the other fades out (called a dissolve) and you can't have one scene slide onto the screen as the other slides off the screen (called a wipe) unless you have some very fancy equipment.

The problem is that you can't mix two video signals together unless they are in perfect sync with each other. They have to be matched up in every detail or the TV just won't know what to do. So how do they do it? They use something called a sync generator. A sync generator is a device that does nothing more than produce a nice clean sync signal that is shared by all your equipment. The catch is that your equipment must be able to accept external sync from the sync generator in the first place. Let's say that the equipment can accept external sync. In this case the cameras, switcher/fader, and other things are all working off the same sync signal. You can now do dissolves and wipes from one camera to another with no problems at all. But there is a different problem when you start talking about VCRs. VCRs march to their own drummer called the control track. Their rhythm is also a little off which means that even if you could get a VCR in step with everything else it wouldn't stay in step very long.

There are three ways to get a VCR in step with the rest of your equipment. First, have the VCR adjust itself. This requires a VCR with a capstan servo mechanism that will accept an external sync signal. The capstan servo mechanism listens to the external sync and then speeds up or slows down the VCR keeping it in sync. This is simple, easy, and very expensive.

**Time base corrector**

The second way to solve the mis-matched sync problem is to let the VCR play along at its own rate and then match up the signal after the fact. This is a bit more complicated but is less expensive. You take the signal from the VCR and pass it through a TBC (time base corrector). The TBC is a kind of buffer that stores the video signal coming in and then passes it out at a nice clean rate. Imagine the video signal coming from a VCR as a crowd of people getting off a subway. Since the subways don't always run on time the crowds grow and thin out as the VCR speeds up and slows down. A TBC acts like a turnstile. It only lets the people through at a certain rate. If you can get the TBC in step with the external sync everything is fine. As long as the VCR is pretty close to regular, the TBC can adjust but when the VCR varies too much the TBC can't correct things any more (and can even make things worse)!

Professional VCRs can produce an almost steady signal and only need a little correction from a TBC but home VCRs are much sloppier. Their signals are so erratic that you would need to correct almost every frame as it came along. Right now, professional TBCs only correct about 8 lines of a single field (a timing error no greater than 1/2000 second) and cost between $6,000 and $20,000. To correct a typical home VHS VCR signal you pretty much have to correct every frame requiring a TBC that will store up to 520 lines of video, these units are sometimes called a Frame Store TBC. Again, this is an expensive solution.

**Genlock**

The third way around the problem is to match the rest of your equipment to the VCR and let the VCR play along at it's own rate. The device used to match a piece of equipment to an external sync source is called a genlock. If the VCR's sync varies then everything that is genlocked to that sync will vary too. That is what a genlock does.

Let's say you have a VCR sending video and sync downstream. You want to be able to mix the video from a camera or other device with the video from the VCR. You feed the signals from the VCR and the camera into a genlock. The genlock "listens" to the sync coming from the VCR and matches the camera sync to the VCR's sync. Simple, effective, and not too expensive. While a genlock isn't a universal solution for every video device it does work for the Amigas.

Genlocks for the Amiga usually have at least two input ports (one for the Amiga and one for an external source like a camera) and at least one output. When you want to mix video from the Amiga with another device the first thing you have to do is get the Amiga signal cleaned up and converted to color composite with an encoder. Fortunately, all the genlocks available for the A500 and A2000s have built-in encoders. Now, you feed video from the Amiga into one port on a genlock device, and the video from another source into the other port. The genlock takes the video from the Amiga and passes it through its built in encoder to get a nice clean signal. It then matches the Amiga sync with the sync coming from the other device. Now that the Amiga is in sync, the two video sources can be mixed together.

## Not All Genlocks are Created Equal

You are walking around your local Amiga dealer and you see a camera on a tripod. Behind it there is an Amiga with a monitor. When you walk closer your live image appears on the screen with Amiga graphics superimposed right on your face. Since you have already read this chapter you know that you are seeing a genlock demonstration. The screen looks great. Nice sharp image, clean lines, wonderful color. You decide that you need a genlock so you buy it and take it home. After you read the manual and hook it up, sure enough the image looks great on the screen. You make a recording of the genlocked image, do a little editing, and voila! instant desktop video. But wait a second. When you play back the edited tape the colors look off. The reds are spilling all over the screen and the blues are nonexistent. What happened?

What happened is that you got a lousy genlock. Yes it can let you superimpose graphics on top of video and the picture looks just fine on a monitor, but the video signal is all wrong. Most monitors and TVs are very forgiving. Video tape isn't as forgiving. If you start out with a poor video signal, it gets worse much quicker than normal when you make copies. Normal VHS dubs lose about 10% but if you start with a bad signal each dub might lose 20% to 50%. So, how can you tell a good genlock from a bad one when you see it? The answer is, you can't. There is no way that you can see the quality of a video signal by just looking at the picture on a normal monitor. You must run the signal through a waveform monitor and vectorscope to analyze the video signal electronically. But most people don't have these specialized pieces of test equipment. So what do you do? Simple, you just keep reading.

At this moment I know of nine genlocks available for the Amiga at nine different prices. I am sure there will be more made and I am also sure that some people will argue with what I am about to say, but you have to start somewhere. Based on test results that I have seen, people I have talked to and articles I have read, I can only recommend three (possibly four) genlocks in three price ranges. Like everything, you get what you pay for so decide how serious you are, what your application is and how much you can spend.

## ProGEN from Progressive Peripherals and Software          $449.95

The ProGEN is a very simple unit to hook up and operate. Simply plug in the RGB cable to the "monitor" port on the Amiga, plug your monitor cable into the ProGEN, the video in (from your VCR or camera) into one port and the video out port feeds the final product to your master VCR. There are no controls on the unit itself so everything is handled through software.

Once you have the unit installed you can either run the software directly (by clicking on the icons from Workbench) or by using the CLI. The software controls the background (external video) and graphics (Amiga generated video) fade rates.

While there are other genlocks available at a lower cost the ProGEN will give you about the best quality video for under $500. It is not quite broadcast quality but if you are on a tight budget then this unit is about the best for the money.

## SuperGEN from Digital Creations          $749.95

The SuperGEN is an ideal genlock for the desktop video situation. The unit is very easy to install and operate. While there are dual sliders on the unit to control the fade rates of background and graphics by hand, you can also control these features with a joystick or through software for nice smooth transitions. SuperGEN gives a very good signal and provides a few extras for the more serious user. Not only do you have the standard RGB in and out ports and video in port, there are two

NTSC out ports for monitoring and recording the genlocked signal. A video thru port (for monitoring the external video signal untouched) and for the more advanced users there is a key out port for use with a switcher/fader.

SuperGEN does come with software that will place your Amiga into interlaced mode, generate NTSC color bars for calibration, even let you use a joystick to manipulate your fades. It is simple, powerful, gives a very good picture and is inexpensive.

**Magni 4004 from Magni Systems Inc.** $1695.00

The Magni genlock is probably the most popular genlock for serious desktop video producers as well as cable, industrial and broadcast users. The video signal it produces is about as perfect as you can get. It has all the features you would expect from a professional quality genlock and more. Of the three genlocks here it is the hardest to install simply because it comes as a two board unit for internal installation (which means it will only work with the 2000 series Amigas). Controls are either through software or with an optional remote control box ($300 extra). Since the Magni uses two internal slots (one of them the video slot) there are compatibility problems with some hardware products (most notably the Flicker Fixer which also uses the video slot) but if you can afford the Magni and can live with the flicker then this genlock is at the top of the list.

## Other Genlocks

**Image Master Amiga Pro-Genlock from Neriki/Telmak** $2200.00

Everything about the Image Master is geared toward the professional user including its rack mountable case and the quality of the output (not quite Magni quality but definitely up to broadcast standards). There are, however, a few qualifiers about the Neriki unit that might make it unsuitable for desktop video. The price is a big factor, also the unit requires an external sync source (not a problem in a studio but could be a problem in a home situation), and there have been some questions about the quality of the components used. Neriki is planning to come out with two new versions of their genlock in late '89 or 1990, one of them a more "home" version at half the price.

**Gen/One from Communications Specialties Inc.**      $849.95

I have heard that this unit provides a pretty good signal, has adjustments for the RGB components of the Amiga signal and has a few other professional features. The main drawback is a short RGB cable making it awkward to use in a studio environment.

**Scanlock from Vidtech Corp.**      $995.00

I have seen the output of other units made by some of the same people who make the Scanlock and they were just about perfect so I would assume that the Scanlock also has a good output.

There are other genlocks out there but word on the street is that they aren't that great. The Neriki, Gen/One and Scanlock are all supposed to be good units but you should test them thoroughly before buying. The ProGEN, SuperGEN and Magni units will give you good to great output in three price ranges. For a few extra dollars most companies offer genlocks designed for use with S-VHS equipment and many companies make PAL versions of their genlocks. SECAM users will have to shop around a bit. Remember to buy the genlock that matches your system and your country. A PAL genlock will not work in the U.S. and an NTSC genlock will degrade an S-VHS signal.

You should also be aware that genlocks are sensitive units and electronic components do go bad occasionally so test your genlock when you get it and if there is a problem return it right away.

## Hooking up a Genlock

So once you made your decision on a genlock the next step is not hooking it up. The next step is reading the manual until you understand everything (or at least as much as you can). Genlocks, like all electronic equipment, are sensitive beasts and can easily be ruined if they are hooked up incorrectly. The ProGEN and SuperGen units will work with any Amiga and connect to the RGB out port on the back of the computer. The Magni genlock only works with the Amiga 2000 series and must be installed internally. It is very important to follow the installation instructions carefully. Don't force any connections, make sure that the boards and plugs are seated firmly, and avoid static at all costs (a small static spark can destroy computer chips.) Once it is hooked up you shouldn't have to worry about it again.

Once the Genlock is connected to the RGB out of the Amiga (the Magni does this internally) you connect your Amiga monitor cable to the RGB out port on the genlock. Depending on the genlock your Amiga monitor will probably only show you the Amiga image not the complete genlocked image. All genlocks have an input for the external video usually called "video in". You can connect any NTSC video source to this input including cameras, VCRs or even another encoded Amiga signal. It is the genlock's job to match the Amiga video to the incoming video. All genlocks will also have a port for the final genlocked video usually called "video out" (the SuperGen calls it "Overlay 1" and "Overlay 2").

The fancier genlocks will sometimes provide additional ports like video thru, preview, video reference in, key out, black burst, external key in, sometimes separate RGB out, etc. Video thru passes the external video thru the genlock untouched. Preview lets you monitor the genlocked image before it is sent downstream. Video reference in can be either an external video signal or a sync generator signal. Key out, black burst, and external key in are all ports for connecting the genlock to a fader/switcher and/or SEG (special effects generator).

**BNC**

Many genlocks also use BNC (Bayonet Nut Connector or British Nut Connector) plugs rather than RCA phono plugs. You can buy BNC to Phono adapters at any Radio Shack. The Amiga/genlock should be downstream from your source or slave unit feeding the final genlocked signal downstream to the master unit. Figures 8 and 9 show the simplest ways to connect a genlock. While all genlocks for the Amiga have a built-in encoder, most of them work better if they have external sync to work with so even if you aren't planning to superimpose Amiga graphics onto another video image it is still a good idea to feed an external signal to the genlock. Try to use as stable a signal as you can. A video camera with the lens cap left on will give a better signal than a VCR. Also, you won't have to worry about the tape ending right in the middle of an important transfer. Just set the background (external video) to zero percent and the overlay graphics (Amiga generated images) to 100 percent.

Many genlocks will also have additional switches, controls and adjustments, and you will have to read the manuals to find out exactly how to use them. The general rule is to read the manual carefully, switches and dials that are difficult to get to (inside or hidden) shouldn't be touched unless you know what you are doing, all other controls are fair game for experimentation but you should make a note of their factory settings so you can go back to them later.

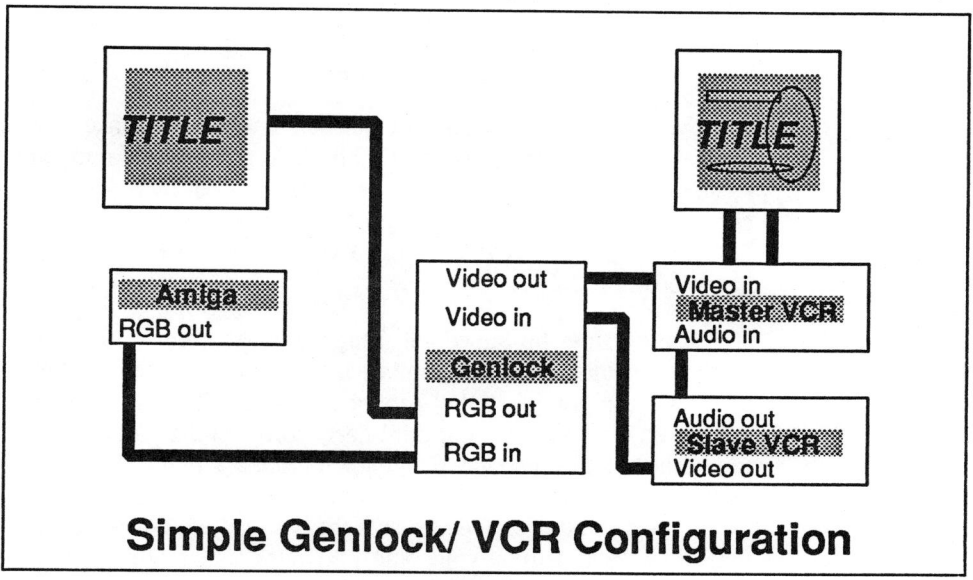

*Figure 8*  *Simple Genlock/VCR Configuration*

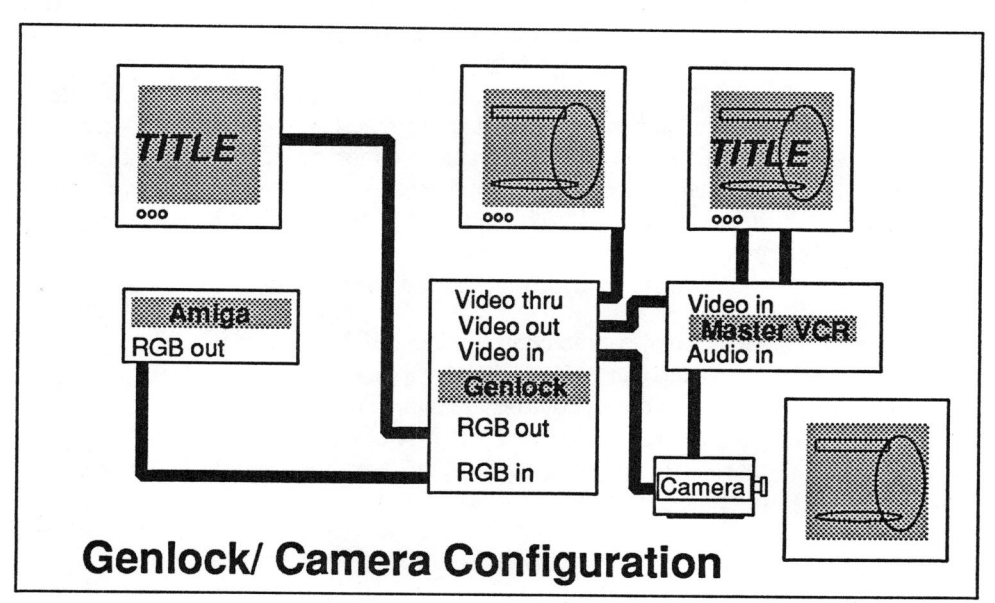

*Figure 9*  *Genlock/Camera Configuration*

## Using the Genlock

Once you have your genlock all connected how do you make it work? Most genlocks get their power from the Amiga so there is no On/Off switch. When the computer is on the genlock is on. The next step is controlling the genlock. There are three ways to control a genlock. Directly, with the software supplied with the genlock, or with third-party software.

If you are using the SuperGen, Magni with remote control box, or another genlock with external controls you can operate the unit independently of what the Amiga is doing. This is very handy, particularly if you are using third-party software that does not multi-task (the ability to run more than one program at a time). Whether your genlock uses sliders or dials you can usually adjust the amount of external video, Amiga video, or both that will be sent downstream. The fancier genlocks will let you control the rates manually or you can set the rates for automatic dissolves. Basically, what you are doing is adjusting the amount of video from the external source that will be sent downstream (from zero to 100%) and the amount of Amiga video (from zero to 100%) that will be mixed in. If external video is set to zero and Amiga video (or graphics) is set to 100 then only the Amiga video is sent downstream. If the external is set to 100 and the Amiga video set to zero then only the external video is sent downstream (no graphics at all). To slowly bring up graphics on top of video you would start with external set to 100 and Amiga set to zero then slowly move the slider (or dial) up until both are at 100% (don't worry about providing too much signal the genlock should automatically compensate for you. It should never send too much signal downstream).

The second way to control a genlock is with the software supplied by the manufacturer. You will have to read the manual carefully to find out what the various controls are and what they do. At the basic level, the software lets you control the mix of external video and Amiga video that is sent downstream just like an external control unit does. Most of the time the software will let you control the dissolve rates (how fast a transition) and the saturation (how much signal) for the two video sources.

Many times the software will use "hot keys." Hot keys are standard Amiga keys that trigger a specific action when pressed. Usually the function keys are used to perform these actions. For example, one action might be to instantly superimpose Amiga graphics on the incoming video (100% external and 100% Amiga). If this action is assigned to Function key one (<F1>), then whenever you press <F1> the software instantly sends the proper commands to the genlock and the action is performed.

Usually the software that controls the genlock can be run as a background task which means that you can have other software running at the same time (a paint program or titling program for example). Even though the Amiga screen doesn't show the genlock software the hot keys are usually still active. Let's say you have the genlock software running as a background task (with hot keys) and at the same time you have a titling package putting titles on the screen. In the above example you could still press <F1> and the genlock software would send the proper commands to the genlock without disturbing the titling program. Since many video and graphic programs use hot keys and sometimes two pieces of software will use the same keys, most genlock software will let you change which keys are used as hot keys. This is called assigning or mapping the keyboard. Check the manual for the proper way to do this.

**BACKUP COPY**

Software control of the genlock will vary from one unit to the next so read the manual and experiment. Remember that there is nothing that you can type on a computer keyboard that will hurt the computer. You can, however, hurt the software on disk with the wrong commands so ALWAYS USE A BACKUP COPY OF YOUR SOFTWARE!

The final way to control your genlock is through third-party software. Some of the video titling and graphics software for the Amiga can control some genlocks. The only way to be sure that the software will control your particular genlock is to check on the package or contact the software manufacturer directly. For example, Broadcast Titler from InnoVision Technology has special software that will control the SuperGen. Just keep in mind that unless the software specifically states that it will control your particular genlock you will probably have to use the genlock's software or external controls.

## Summary

Since the genlock is one of the key elements of a desktop video studio, I will be talking about them throughout the book. This chapter is just to get you familiar with what a genlock does, how it operates and how to connect one. The actual uses of the genlock will be covered in the chapters that deal with specific applications like titling, paint programs and special effects.

# Chapter 4

## Digitizers and Scanners

# Chapter 4
# Digitizers and Scanners

The next weapon in the desktop video arsenal is a digitizer or scanner. Basically, these devices will take an image and turn it into something that the computer can understand. Once you have it in the computer you can store it on diskette, load it into a paint program, modify it electronically and then, using a genlock or encoder, send it out as video information.

A digitizer gets the image from a video source, usually a camera, while a scanner uses a device like a photo-copier. The biggest difference between the two is that a scanner can only take an image from a two-dimensional source (like a piece of paper or photograph) while a digitizer can take any video image. There are lots of varieties of digitizers and scanners and we'll take a look at some of them. But first a little background on why and how they work.

## Why Digitize?

Computers are basically a bunch of on and off switches put together in a very complex pattern in a small box. Even though computers can do amazing things deep down inside they can only understand two things; on and off. A clever designer or programmer can make the computer behave as if it can understand more than that, but when you really analyze the situation you find that computers live in a digital world while people live in an analog world. When we want the computer to do something with information we have to break the information down to a series of ons and offs. Once we do that the computer is happy.

So how do they get the computer to generate all those wonderful images? It is just a very large, very complicated series of ons and offs. If we want the computer to take a picture of something we have to break that picture into a series of ons and offs. This is what a digitizer does. It takes an image and turns it into a string of ons and offs so that the computer can deal with it. A digitizer is essentially an A to D (analog to digital) converter.

If computers had unlimited memory (which they don't) we could get them to break down a video image into so many ons and offs that you wouldn't be able to tell the computer image from a real one. With the right hardware and software you can get close to that point but there will always be a little graininess or pixelization. A pixel is the smallest dot on a monitor that the computer can manipulate and the highest resolution that an Amiga computer can produce is something less than what a normal television set can.

It isn't really a matter of not being able to display that amount of information, it is more a question of how fast the computer can do it. If you break down a normal video frame into ons and offs it turns out to be about one mega-byte of information. Multiply that by 30 frames a second and the computer just throws up it's electronic hands in frustration. A computer can't generate that amount of information that quickly. But most of the time the computer doesn't have to. If all you are doing is putting text on the screen or a few simple graphics that don't change very rapidly the computer is more than adequate. After all computers are built mainly for manipulating information not manipulating picture tubes.

So a digitizer takes a video signal and translates it for the computer. Since computers don't usually deal with pictures that are as detailed as a video image, the digitizer also simplifies the image. If there are four pixels in a block and three of the pixels are green, the digitizer tells the computer that there is only one block of green. Many digitizers will average the pixel colors, trying to approximate the color.

**Interchange File Format**

Since the computer also stores and manipulates information differently than video equipment does the digitizer software performs one more function. It takes the digitized screen image and converts it into a standard graphics format. The Amiga can store and display graphics in a number of different formats and resolutions but one of the standard formats for saving pictures to disk is called IFF (Interchange File Format). Most paint programs for the Amiga use the IFF picture format. All you really have to know about IFF is that it is just a way that the computer stores graphic information on diskettes. Once a digitizer or scanner captures an image you can usually save the image to disk as an IFF file.

## Types of Digitizers

Digitizers can be broken down into any number of categories but the two that make the most sense are real-time digitizers and digitizers that take a few moments to operate. There are a few digitizers that can capture an image in as little as 1/30th of a second (although they still take a second or two to process the image after it is captured). These units could also be considered frame grabbers with digitizing capabilities (I'll talk about frame grabbers in Chapter 5). They are usually a bit more expensive than non-real-time digitizers. You should also know that when operated at the fastest speed the image they produce is not as high quality. If you want to digitize images from video tape or any moving subject then a real-time digitizer is what you need. Real-time digitizers usually include a special slow-digitizing mode for static images that will increase the quality of the final digitized image.

**Non-real-time digitizers**

Non-real-time digitizers are usually less expensive than their real-time cousins and they can usually produce a better image. One of the ways that they do this is by spending more time during the digitizing process for image processing. The other trick used by digitizers is that they usually require using a black and white camera. Black and white cameras have a higher resolution and give a much better picture to begin with (this is why it is better to have a black and white viewfinder on your camcorder).

The digitizers that use a B/W camera use an old photography trick for producing color. By scanning a black and white image three times through a series of three colored filters (red, green and blue naturally enough) and then combining the three pictures you get color.

**Image processing**

Image processing is a computer term that means using the computer to enhance, modify, even guess what the final picture should look like.

The simplest form of image processing is adjusting the saturation, hue, tint, contrast and brightness of the image. In its more advanced stages image processing can include things like dithering, anti-ailiasing and color averaging. Dithering is a process where two colors are mixed to produce a new apparent color, a little like an optical illusion. Anti-ailiasing is a video term adopted by computer people. When you have two shapes of different colors on a screen the edge can appear ragged or the colors can slide or overlap producing a false color. Anti-ailiasing looks at the two colors and places a border between them that is a half-way color. Color averaging is when you may have ten or twenty shades of blue in a sky but only wish to use one or two. The computer decides what the average color blue is and picks the best shade for you.

At the very high end, image processing includes things like computer colorization of old black and white films or NASA image enhancement of the Voyager flight pictures.

## Scanners

If you want the ultimate in digitizing resolution then you have to move up to a scanner. Scanners are devices that are, in a fashion, dedicated digitizers. Like a photo-copier, a scanner uses optical technology to scan a two dimensional image. High intensity light is shined on a two dimensional subject a single row at a time. The light is measured and fed into the computer line by line. The advantage of a scanner over a standard digitizer is that you can attain a much higher resolution density, actually higher than the computer can reproduce. This is very advantageous if you are working in a print medium which is why scanners are most often used in the publishing industries.

The disadvantages of a scanner are that an image can have so much detail that it uses much more memory than can be installed in an Amiga. In fact, a single 8 X 10 black and white image scanned at 300 DPI (dots per inch) will consume about 900,000 bytes of storage space (an Amiga diskette can only store about 880,000 bytes). Scan the same image in color this time and it now takes about 21,600,000 bytes (20 megabytes) to store that image. When you get into these realms it takes more than a simple paint program to modify the image and more than a simple printer to get a hardcopy. Now, you don't have to scan images at that resolution or size. You can set the scanner so that it gives you only the same picture clarity of a normal digitizer if you want. You can also scan a small part of the total image, filling only a portion of the screen. The other major drawback to scanners is their price. Expect to pay in the thousands for a good scanner and the hardware necessary to use it.

If you already have a scanner or have a need for one there are definitely ways that you can use a scanner in desktop video. Hand drawn materials will transfer to video much clearer than by simply pointing a video camera at the canvas or paper. A scanned image of an actor in a corner of the screen next to their name makes credits more interesting visually. Montage work is much easier to do if you have still photographs before you to compose from, rather than trying to find pieces of video taped in a thousand locations. If you need a three dimensional image you can always take a photograph and then scan the photograph. You will find that many times you can get a much cleaner and sharper still image with a photo camera and a scanner than with a camcorder. And finally, broadcast studios have spent thousands of dollars on equipment that will pixelize a screen, an effect that can be reproduced easily with a scanner or digitizer. The special effects ideas are unlimited.

If you are looking to buy a scanner then the choice right now is fairly easy. Professional ScanLab (PSL) from ASDG is one of the best scanner interface/controllers available for the Amiga. Used in conjunction with a Sharp Electronics Corp, JX-450 or JX-300 color scanner PSL can give you larger, more detailed images than the Amiga can display when you need it to, but it will let you select parts of an image, manipulate the colors, aspect ratios and resolution in an almost unlimited number of ways. It only works with the A2000 series (the interface is all on one board) but the results are very, very nice.

## Selecting a Digitizer

When you start talking about digitizers for the Amiga there are a few units out there but there is one unit that has completely dominated; Digi-View from NewTek. In fact it has so dominated the Amiga market that other digitizers have had to either drop out of the digitizer market or upgrade their units to the frame grabber category. While you might feel that NewTek having a monopoly in the digitizer field is limiting, or that they might not bother to improve the Digi-View unit, it does make selecting a digitizer very easy. And even though they didn't really have to improve things NewTek has released improved versions of Digi-View almost every year. The current version, Digi-View Gold 3.0 is by far the best digitizer and software at that price. While it is not real-time, and it requires a black and white camera (or a device called a color splitter) the results are impressive. So, if you want a digitizer (and you probably will) the choice is easy, just buy Digi-View Gold.

Another advantage of Digi-View is that since it is the Amiga standard in digitizers, just about every other piece of software that deals with digitized images will work with Digi-View. NewTek also sells accessories for Digi-View including black and white cameras, copy stands with lights and Digi-Droid a motorized filter wheel unit that is computer controlled. (NewTek also produces a HAM paint program called Digi-Paint, particularly well suited to video work it is capable of producing graphics in any Amiga resolution).

## Hooking up a Digitizer

Since most digitizers take at least a few seconds to digitize an image, and getting the kinds of results that you want are a matter of trial and error it is best to set up your digitizing off-line. In other words you probably will not be digitizing images during your taping or dubbing. Digitizing is a process that should be done separately and then you can incorporate the results later.

The simplest set up is a black and white camera (depending on the digitizer you use) on a stand or tripod, the colored filters or filter wheel mounted on the camera, then the camera feeding directly into the digitizer. Digitizers vary on how they are connected to the Amiga. Digi-View connects directly to the parallel port (where the printer would normally connect). As usual, with any piece of equipment you should read the manual carefully before you connect anything. (See Figure 10.)

If you are using a black and white camera you will probably want to first connect the camera directly to a monitor for focusing and once the camera is set then connect it to the digitizer. If you are using a camcorder or camera with auto focus you may wish to turn off the auto focus to conserve power and avoid vibrations. Your camera should be mounted as solidly as you can manage because any movement during the digitizing process can distort the final image. You should also try to get the filter wheel as close to the lens as possible without touching it. The wheel should turn freely to avoid moving the camera when changing to the next color.

**Color Splitter**  If you want to use a color camera or camcorder you will have to buy a color splitter. This is a small, simple unit that connects between a color camera and the digitizer. Basically, a color splitter acts like the filter wheel on a black and white camera sending in turn a red only, green only, blue only video image to the digitizer. SunRize Industries makes a color splitter that works well with the Digi-View digitizer. It also has separate adjustments for saturation and hue. (See Figure 11.) The advantages of using a color splitter are that you don't have to buy a separate black and white camera and by flipping a switch on a remote unit you eliminate the chances of moving the camera while turning a color filter wheel mounted on the camera. The disadvantage to using a color splitter is that color cameras don't usually give quite as sharp an image as a black and white camera. If, however, you have a camera that outputs RGB separately you will get a much better image.

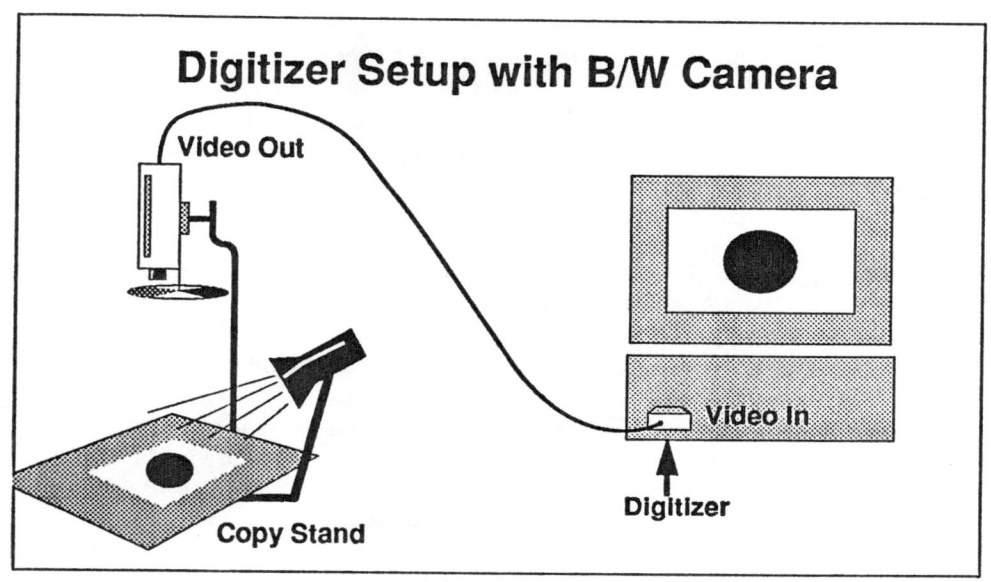

*Figure 10*  Digitizer Setup with B/W Camera

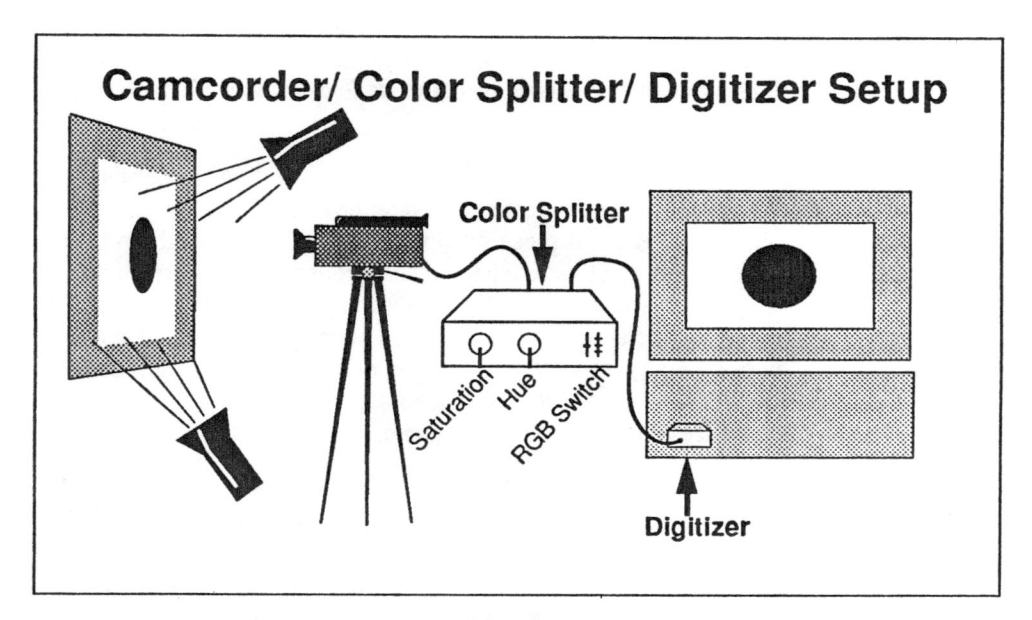

*Figure 11*  Camcorder/Color Splitter/Digitizer Setup

Another problem you may run into if you are using a camcorder is that unless the camcorder is actually taping many of them will automatically shut themselves off after a few minutes. One thing you might try is removing the tape from the camcorder. You will probably get a warning light flashing in the viewfinder but the camcorder won't shut itself off. Since digitizing takes a fair amount of time (with tests, trials and experimentation) it is a very good idea to use an AC power adapter for your camcorder rather than the batteries.

### Lighting

One of the most critical aspects of digitizing is proper lighting. It is worth while spending the time to set up proper lighting for your copystand or subject area. The Digi-View manual suggests using fluorescent lights rather than incandescent lights because of the flicker rate, but whatever type of lighting you use you will probably have to experiment. Cool white fluorescent lights seem to work the best. When adjusting the lighting and focus you can use the viewfinder (if your camera has one) but feeding the video into a full sized monitor will give you a much better idea particularly concerning glare and 'hot spots.' One trick to test for hot spots is to digitize a solid grey piece of cardboard. If you plan to do a lot of digitizing you might wish to pick up a color separation guide and grey scale reference chart at a local camera store. Paste the two charts on a stiff piece of cardboard and use it at the beginning of each session. This is the same idea as color bars used to calibrate video cameras. You can use the lowest resolution, fastest scan rates to adjust the colors in Digi-View before going to your main digitizing.

If you are planning to digitize a number of two dimensional images in one session you might consider using the side of a refrigerator with a thin neutral colored piece of cloth or paper as a background (black velvet is the ideal material for a background as it absorbs light). By using ordinary magnets you can easily and quickly change and frame the pictures you plan to digitize. You can keep paper objects nice and flat by placing a sheet of glass over them during the digitizing process but you will have to be extra careful about glare and ordinary window glass will cause colors to shift toward green.

### Burn in

A few other points that you need to know about digitizing are that it is not a good idea to let your camera remain pointed at the same image for too long. Even though modern cameras are less susceptible all cameras can suffer from 'burn in'. Burn in is when an image (usually a bright image) gets 'stuck' in the camera so you can see a ghost even when the camera is looking at something else. It is a good idea to put the lens cap back on between each digitization. You also need to know that video is an interlaced medium and while you can change an image after the fact (with the right software) it is easier to digitize your images in interlaced mode to begin with.

## Digitizing Process

The digitizing process is very simple once everything is set up and your manuals should explain all the subtitles. The basic steps are:

1. Turn off the computer and plug in the digitizer.
2. Set up the camera, lights and subject to be digitized.
3. Connect the camera to a monitor so you can check the camera focus and adjust the lights for hot spots.
4. Once the video image is the way you want connect the camera video out to the digitizer.
5. Turn on the computer and load the software.
6. Determine the resolution and digitizing mode you will be using.
7. Turn the filter wheel so that the red filter is directly in front of the camera lens and activate the red scan function on the computer. (Or set your color splitter to the 'red' setting, if you are using a color camera/color splitter combination.)
8. When the scan is finished turn the filter wheel to green and activate the green scan. (Or set your color splitter to the 'green' setting, if you are using a color camera/color splitter combination.)
9. When that scan is finished turn the filter wheel to the blue filter and activate the blue scan. (Or set your color splitter to the 'blue' setting, if you are using a color camera/color splitter combination.)
10. When the final scan is finished put the lens cap back on the camera.
11. Activate the "display" function and your final digitized image will appear.

From there you can go back and adjust the color balance, sharpness, contrast, etc. When you get an image that you are satisfied with save the image to disk and you are ready to digitize the next subject.

## Summary

While it is possible to get very clear images with a digitizer you will probably end up using one for special effects. By moving the subject material or camera during the process, changing the order of the filters or mis-adjusting the color settings on purpose you can create dozens of interesting effects while still retaining a recognizable image. If you aren't that good an artist you can digitize an image and then bring it into a paint program and make it look hand drawn. Once you have an image digitized you can change the colors, size, even animate the image with the right software. A digitizer can open up an entire world of options for titling and special effects and the cost of a good digitizer is so small that you will probably want to pick one up sooner or later.

# Chapter 5

## Frame Grabbers
## Frame Buffers

# Chapter 5
# Frame Grabbers
# Frame Buffers

If the idea of turning color wheels, avoiding vibrations or waiting for a digitizer to process an image is unappealing, or if you want to capture an image in real-time from a video tape or off the air then you have to move up to the next category of video digitizers called frame grabbers or frame buffers. A frame grabber captures a video image in as short as 1/60th of a second. Once the image has been 'grabbed' and then digitized you can then save the image to disk, manipulate the image with a paint program, modify the image and eventually put it back out to video tape with a genlock or encoder. Technically a frame buffer is a device that can hold more information than the computer can display either for import or export. Most of the time, for video applications, you would be using a frame buffer as a frame grabber. So most of the time I will use the term frame grabber to designate both units. But I'll talk about the differences in more detail a bit later.

Like a digitizer, the frame grabber is an analog to a digital converter. Computers may be very fast when it comes to processing information but they can only input information at a certain rate (much slower than a video signal outputs information) so the first function of a frame grabber is to act as a temporary storage device for that video information. Since computers only understand ons and offs the frame grabber takes the stored video information and converts it into small pieces of digital information. Computers don't usually display the vast amount of information contained in a standard video image so the frame grabber also performs some simplification of the information. Finally, the frame grabber software will let you store the information in a number of formats, with IFF being one of them. That way you can later load the picture into a paint program or an image processing program for modification.

**Frame store units**

Frame grabbers and frame buffers or frame store units are not new to the world of video. Some broadcast formats (like quad tape) cannot display a freeze frame or operate in slow motion. A frame store device would collect the information until it had a full frame of video and then display it. If you wanted to show slow motion the devices would collect frames and send them back out one at a time. A frame grabber is essentially a video storage device.

When you start talking about computers and video the frame grabber is more than a storage device. The computer is capable of holding all the information contained in a frame of video but it can't gather that information as fast as a video signal sends it out. The frame grabber acts as a quick input device, gathering the information as fast as the video comes in and then holding it until the computer can digest it all. A non-real-time digitizer doesn't have to store a lot of information because it operates at the speed of the computer. Feeding the information at the rate that the computer can deal with it. That is why the digitizers, I talked about in Chapter 4, take so long to digitize an image.

A frame grabber, on the other hand, must have a certain amount of internal memory (RAM) to hold the information temporarily until the computer can deal with it. This is the main reason that frame grabbers (or real-time digitizers) cost more than non-real-time digitizers and involve quite a bit more electronic circuitry.

## Real-time digitizers, Frame Grabbers and Frame Buffers

There are three varieties of real-time digitizing devices available for the Amiga. Real-time Digitizers, Frame Grabbers and Frame Buffers. They will all take a video image and digitize it but there are some differences in how they operate. Right now there are one or two real-time digitizers for the Amiga. They perform the same functions that a non-real-time digitizer does only much faster. One unit will only digitize a black and white image in real time but requires using filters to do a color image. Another unit does digitize color images in real-time but lacks resolution and clarity.

At the moment there are a few frame grabbers available for the Amiga that can grab an image in real-time but they take a moment or two to process the information. The biggest difference between a frame grabber and a digitizer is that a frame grabber gathers all the information in one quick bust and then holds it for a moment while the computer catches up. Since frame grabbers operate in 1/30th of a second (for a color image) or 1/60th of a second (for black and white) you can grab an image from video tape, a video camera, even off the air.

**Interfield Jitter**

Frame grabbers are susceptible to a problem not encountered with other digitizers. Since they grab a frame and not a field you will occasionally get jittering. Particularly when there is fast action in the source video one field will vary from the next. If you are playing back the fields at 60 times per second the human eye can't tell that some of the fields don't match up exactly. However, when you freeze one of those frames containing two mismatched fields you can easily see the jittering. This is called, naturally enough, **Interfield Jitter.** Some frame grabbers use software or hardware image processing to reduce this jittering but the best solution is to simply try again. Another problem that frame grabbers suffer from is that they need a fairly good, stable signal or they just won't work.

Right now there are two frame buffers for the Amiga. The biggest difference between a frame buffer and a frame grabber is that a frame buffer is a two-way device. While a frame buffer has features that let it perform the duties of a frame grabber-digitizer it can also store information that the computer generates and then output it to another device that can use it. The Amiga computer is only capable of displaying a certain amount of information using a certain amount of colors at a certain resolution. A frame buffer expands this range considerably. The frame buffers available for the Amiga now are capable of dealing with up to 16.7 million colors at a resolution of 746 X 484. That is much more than an Amiga can deal with. So why would anyone want or need such capabilities? There are two reasons why you might want a frame buffer over a frame grabber. In very high-end image processing and graphics work there are computers, scanners, printers and "graphic engines" capable of dealing with much more information than the Amiga can display. Some graphic programs for the Amiga can also output that kind of information (even though you can't see it displayed on an Amiga screen). The main applications in this area are CAD (computer aided design) and publishing. The second reason for using a frame buffer is that you can store a video image in the buffer (or store it on disk) for later use in video without losing resolution. The act of digitizing an image for use in the computer means degrading the image somewhat. If, however, you keep the image in its original form you can still output that image to video without losing as much. You will still lose some quality and you won't be able to see the results until it goes back to video but it will be much better than a normal digitized image.

The frame buffer can be "loaded up" from either the Amiga side or from the video side. When used as a frame grabber/digitizer the video image is stored temporarily then processed so that the Amiga can display it. When used as a frame buffer you can either have the Amiga software slowly build an image in the buffer or simply hold a video image for later export to another device like an ultra-high resolution printer, color separation device or even back out to video.

## Selecting a Real-Time Digitizer

**LIVE! from A-Squared**
$299.95 (A1000), $399.95 (A500), $450.00 (A2000)

A-Squared manufacturers LIVE! for the Amiga; a real-time digitizer. LIVE! has gone through a number of incarnations, owners, manufacturers, problems and successes. LIVE! was the first digitizer for the Amiga displayed at the computer's launch more than three years ago. Originally designed to digitize images as fast as possible the unit cuts some corners when it comes to resolution and picture quality. Some experts claim that LIVE! is really a bit-plane grabber because of the unique way that it works. Even though the quality may not be as high as other units if you are looking for real-time special effects the LIVE! unit has some advantages. When used in conjunction with some specialized software from Elan Designs (Elan Performer and Invision Plus) you can do amazing things that just can't be done with other digitizers. Although Elan Performer and Invision Plus were mainly created for live performances, there are a number of digital effects included in their wide range of MTV effects and features.

## Selecting a Frame Grabber

At the moment there are two frame grabbers available for the Amiga and they each perform fairly well.

**Framegrabber from Progressive Peripherals and Software         $699.95**

This external unit has just about every feature you could ask for in a frame grabber. It is simple to install and works in just about every mode and resolution. It can capture a color image in 1/60th of a second (two fields) in non-interlaced mode or 1/30th of a second in interlaced mode but it takes anywhere from 5 to 30 seconds to process the information (depending on the image size and the color settings). On the outside of the unit are controls to adjust intensity, hue and saturation of the incoming video. Operating the unit is very simple (a single key press activates the capture). Another nice feature of the Framegrabber is the ability to switch the display back and forth between the untouched video and the grabbed image. That means no plugging and unplugging cables to adjust camera focus and framing. You can digitize in either 2, 4, 8, 16, 32, 64 or 4096 colors, or you can digitize an image in black and white using a 16 shade gray scale. If you prefer to use a black and white camera on a copy stand with colored filters then Framegrabber will let you do that too (but you won't get quite as good an image as you will with Digi-View Gold).

**ANIM format**  Like other digitizers, once you have the image in the Amiga you can save it out to disk in the standard IFF format but Framegrabber also lets you save images in three other formats. Another feature that you won't find on too many digitizers is an automatic animation/time lapse function. When you set the software in this mode the Framegrabber will automatically grab, digitize and then store images in an IFF ANIM format (a special Amiga file format for animations). This can be done either one frame at a time or automatically at regular intervals (from one image a second to one image every one hundred hours). If done manually, you could do claymation style animations without a lot of special equipment. Just digitize a frame, move your model a little, digitize another frame, and on and on. When you are finished you can play back the animation using any number of ANIM players. In the automatic mode you can do time lapse effects. Point a video camera at the sky, a flower bud, or just about anything that changes slowly and set the rate. The software will automatically grab frames at the specified rate and add them to the ANIM file.

Like other frame grabbers, digitizers and genlocks the quality of the final image depends a great deal on the quality of the video signal being fed into the unit to begin with. Video tape players in the pause mode may not supply a clean enough signal but the advantage of a frame grabber is that you can freeze the action without putting your playback unit in pause.

## Selecting a Frame Buffer

There really isn't much choice here but that may not be a problem. The only frame buffers currently available for the Amiga work pretty well (both as frame grabber/digitizers and as frame buffers).

**VD-1 from Impulse**                                                    **$1000.00**

The VD-1 unit boasts a number of features like those you find on the Framegrabber from Progressive but it is also a frame buffer. It can grab a color image in 1/30th of a second for full frame or 1/60th of a second for a single field. The single field capture is only 200 lines of resolution so don't expect great images but if there is a lot of movement in the scene a single field capture is about the only way to avoid interfield jitter. While the VD-1 only captures information in one full screen, overscanned resolution the supplied software will convert the image to any format. You can save the image as an IFF file or as a full 24 bit RGBN file. This is useful when using the VD-1 as a frame buffer.

The VD-1 is a simple box with NTSC in and out ports and connects to the Amiga via the parallel port. So it is compatible with all Amigas (with at least 1 megabyte of RAM). Rather than toggling between the external video source to the Amiga you would either use a separate NTSC monitor or feed both signals into the Amiga monitor and switch manually. There are no external controls on the unit, all functions are handled through the software. When you wish to capture an image just press a single key. The image is captured, then digitized and displayed through the Amiga. You can then adjust the image (contrast, saturation, hue and dithering) through software and re-digitize the frame. The image is held in the VD-1 until you grab another so there is no need to go back and try again. While the VD-1 does have a time-lapse or animation feature, the images are not stored in an IFF ANIM file. Instead, each image is automatically saved to disk with incremental filenames (IMAGE.1, IMAGE.2, IMAGE.3, etc.) which could later be put into an ANIM file or used by a page flipping program.

As a frame buffer the VD-1 can be 'loaded up' with computer images generated with special software like Turbo Silver or Sculpt 4-D. You can even grab a 24 bit video image for use as a high-resolution background for animations. Since the VD-1 is a frame buffer as well as a frame grabber you can use it as a digital freeze frame device. Holding and even storing images for later output. Included with the VD-1 is a full compliment of software utilities for programmers who wish to create their own software. There is also a paint program for manipulation of the images once they are digitized.

### Framebuffer w/Frame capture from Mimetics $748.00

Sold as two separate units (but functioning as one) the Framebuffer and Frame capture add-on may be a bit of an overkill if all you want to do is digitize a few images. But if you want more than just a few simple special effects then the Framebuffer/Frame Capture system should satisfy your needs. If you also have access to a graphics engine high-end graphics workstation environment then this is a nice solution to the problem of not being able to take the equipment home over the weekend. With the frame buffer and the right software you should be able to not only capture images from a video source you would also be able to create spectacular graphics with the Amiga and display them on another machine. Used as a frame grabber/digitizer the unit works very well but is subject to the same problems that all frame grabbers have. When grabbing a frame with a lot of fast action overlapping fields will cause a jittering of the image, the software that comes with the framebuffer compensates for this somewhat but it is better to just try again with a more stable image. The framebuffer also works better with a solid video source. An internal board configuration means that this is available only for the A2000 series computer so A500 and A1000 owners will have to look for another unit. Used as a frame buffer the unit can either gather information from the Amiga or hold an image for later export.

## Connecting a Frame Grabber

Of the four units discussed only the Framebuffer unit is installed internally, the others are simple boxes with input ports for an external video signal and outputs for sending images out. Most of them intercept the Amiga's signal as it leaves the computer and a second cable goes to the Amiga monitor. The Framegrabber from Progressive lets you toggle the Amiga monitor back and forth between the untouched image to the grabbed image. The VD-1 requires an NTSC monitor. All of them come with supplied software and most of the cables that you will need (the Framegrabber requires an additional cable not supplied with the unit but it is easy to build or it can be ordered separately). Be sure to read all the manuals carefully before connecting any of the units. All the units are activated and manipulated through software (although the Framegrabber has external controls for intensity, hue and saturation).

Like the digitizers, you will want to set up your system off-line. Even though the framegrabbers can grab an image in a fraction of a second it does take a few moments to digitize the images. This means that you won't want to be digitizing during your editing or dubbing. The best idea is to plan your production carefully, decide which scenes you will be digitizing, digitize the scenes and afterward, either put them on tape with an encoder (for editing in later) or, using a genlock, mix the finished images with live or taped video during your editing.

## Frame Grabber Tips

If you have been paying any attention in the last few pages you know that the number one trick for getting the best results from a frame grabber, real-time digitizer or frame buffer unit is to supply the unit with a good solid, stable video image. The first step toward that goal is by using a camera (which will usually give a more stable signal than a VCR). If you are using a video camera then you can still help things quite a bit by making sure your subject is well lit. The next most stable image will be from an off-the-air source like a cable TV or antenna (if the signal is strong and clear). The final source is from a VCR (VCRs are notorious for their unstable signals). Avoid trying to grab an image from a VCR in pause mode, the image is usually unstable, and the point of a frame grabber is to do that very thing for you.

All frame grabbers are subject to interfield jitter problems. While the software can try to compensate for this somewhat they are not perfect. You may get a stable image but it will be blurred. The best solution is to try for a still (or at least a slow moving) image. If you know that you will be grabbing images later you might want to tape a few seconds of still images during your shoot. If you aren't trying to get perfect images you can try grabbing a field in black and white and then artificially adding color afterward or with the VD-1 grab a color field and live with the lower resolution. By scaling the image size down to less than full screen the lower resolution will be less noticeable.

If you are going to be grabbing and digitizing a number of images for a single production some of the units allow you to lock in a color palette. That way the digitizing software will only use those colors when converting the image. This can be important if you are planning to add other computer graphics to the digitized image. The Amiga gives you a wide range of colors to pick from but if your digitized image is using one 32 color palette and your graphic (created with a paint program) is using a different 32 colors either the graphic or the digitized image will shift colors when you try to mix them together.

Expect to spend some time experimenting with the frame grabber. They can be very easy to set up and use but they take some fine-tuning to get the best image.

There is another use for a framegrabber to solve one of the biggest problems in desktop video. With a framegrabber you can simulate a number of AB effects. While it is possible to dissolve and wipe from a VCR to a genlocked camera or a genlocked computer image it isn't possible to dissolve from one VCR image to another VCR image (AB effect) unless the two VCRs are synced together. Unfortunately, you can't sync two VCRs together without some very expensive equipment and special VCRs. You can fade one VCR to black, switch sources, then fade into the second VCR but you can't dissolve straight from one image into the other.

**AB dissolve**

That is where the frame grabber comes in. With a little planning and careful timing you can simulate an AB dissolve using a frame grabber. There are two methods for doing this.

Method 1. During the shooting of your scenes make sure that the end of scene one concludes on a static image. An easy way to do this is have the actors leave the scene or pan the camera away from the action and lock on something else (this is where a tripod is nearly essential). Leave a few seconds for the dissolve effect. Then do an in-camera cut to scene two (providing your camcorder has flying erase heads). Back in the editing room grab a frame at the end of scene one (during the stillness) and digitize it (or hold it if you are using a frame buffer). Use the highest resolution possible and try to get the digitized image as close as possible to the video image. Now, with the original tape in the slave unit feeding into the Amiga (which is genlocked) begin the transfer to the master. Be sure the genlock is at 100% external video and zero% Amiga video. At the end of scene two quickly bring the Amiga video up to 100% and the external video down to zero%. After a moment slowly bring the Amiga video back down to zero% and the external back up to 100%. With careful timing you should be dissolving into the beginning of scene two.

Method 2. This is for dissolving between scenes that are not together on a tape. Again be sure to end a scene on a static image if possible. Back in the editing room transfer scene one to the master. Go back and grab a frame from the end of the scene where there is no motion. You may have to experiment to get the image as close as possible to the original. Once you have the scene digitized in the highest resolution possible set up scene two in the slave unit which is genlocked to the

Amiga. Set the genlock so that it is only showing Amiga graphics (the digitized frame from the end of scene one). Now record a new scene two starting with the genlocked Amiga image dissolving into the original scene two video from the slave VCR. Finally, go back and edit the new scene two onto the end of the original scene one. The final result, if you timed everything correctly, should be scene one coming to an end on a static image which then dissolves smoothly into the beginning of scene two.

Since the digitized image is essentially an Amiga graphic you should be able to do any special effect with it that your genlock or production software is capable of, including fades, wipes, dissolves, rolls, even digital effects. You still won't be able to go directly from one moving scene into another moving scene but a bit of creativity should compensate in just about all circumstances.

## Legal Issues

If you are grabbing images off the air or pictures from a magazine or book then there are a few things you should know. The problem with grabbing images off the air is that 99% of those images are copyrighted. That means that you can use the images for your own private use but you won't be able to sell your tapes, charge people to see them, enter them in any contests or let anyone but your family and closest friends see them. Of course you could always contact the broadcasting company and try to get written permission to use those images. If you are going to try to go that route here is a simple tip for getting permission. Try to contact the public relations or marketing department first. They can be very nice but sometimes hard to get in touch with. If your use is fairly innocuous and in good taste then the odds are that a PR department will give you permission in about 90% of the cases without charging you a dime. They may even help you get other images for free (publicity photos, stock shots, etc.). It might take a few months to get that permission and usually they will just ask that you put a credit at the end of your production saying where the image came from. In the same situation a legal department will only give you permission in about 10% of the cases (after months of thinking about it) and they will probably want you to sign all sorts of papers, demand copies of the final product, perhaps even changing their minds after they see it, and might even charge you a fee for every use. PR people are paid to get publicity for their clients, legal departments are paid to handle things like law suits, liable cases, contracts, copyrights, fees and justifying their existence. If possible you should avoid going through the legal department.

Some companies are notorious for not letting people use their images (Disney Studios in particular). Magazines, newspapers and book publishing companies can also be sticky about images taken from their pages. In most cases it is not that the company is being hard-nosed just to be difficult. It is because they have very rigid contracts with their illustrators and photographers (who all make their livings depending on how many times their work is used). You should also know that even an image you "create" can belong to someone else. If you made a cartoon of a Marilyn Monroe character and that cartoon wins an award in an animation film festival you might expect a call from the lawyers of the Monroe estate. Even if you drew each frame by hand they consider Marlyn's image to be theirs.

Even though most lawyers and judges haven't even heard about digitized images yet you can bet that there will be some confusing court battles ahead. How much of an image must be the same before it infringes on someone else's copyright? What if you substitute a green pixel for every red pixel in a digitized Mickey Mouse? Or substituted different colors for every pixel? What if you modify the image with a paint program? How much would you have to modify it before it is unique?

Right now the only safe pictures you can digitize are from photographs or video tape you shoot yourself (in a public place that is) or images that you have written permission to reproduce.

Of course, if you really, really need a picture of Burt Reynolds in your video you might try sending for an autographed picture (publicity photos are considered public domain) and there are stock photo companies that will track down just about any image you can think of, arrange for all the proper contracts, and then bill you for their troubles.

# Chapter 6

## Software
Paint and Image Processing Programs

# Chapter 6
# Software, Paint and Image Processing Programs

While the hardware is the heart, body and skeleton of a desktop video studio the software is the brain. Unlike video equipment, where you control what happens every step of the way, computers can do nothing without the software. Up to this point I have only hinted at a few programs that are available for use with specific hardware. These next few chapters will be devoted to software that is specially for video applications. If you are new to the computer side of things you might think that the only time you will use a computer is when you want it to do something specific for a video project. Hopefully you will find, once you get used to the computer, that there are thousands of other uses for it. But, that isn't why you bought this book.

## Video Software

If you did a survey two years ago of all the software titles that were for video applications you would have come up with maybe a dozen titles. Now there are over a hundred titles. Now, not all of these titles are just for video. In fact the most important video software wasn't designed for video at all (it is the paint program if you hadn't guessed). Now most people can't just go out and buy every piece of software ever made so I'll make a few suggestions along the way. But first, we should divide the software into categories and describe how each is used in a desktop video situation.

**Paint Programs** A paint program lets you create images on the computer screen using a mouse, the computer keyboard, a graphics tablet or other input device. Whatever input device you use the images created are pretty much free-hand drawings. Paint programs do give you a number of tools to make that drawing process easier. The software will help you draw perfect circles, squares and lines every time. They let you mix colors, undo your mistakes, even add text to your picture. Some paint programs also have a special kind of animating technique called color cycling. They can flip, stretch, distort, copy, move and alter parts of the image with a few key-strokes. Finally, they let you save or load images from disk.

The major drawback in using a paint program for video is that you have to have a certain amount of skill to create a nice image from scratch.

**Image Processing Programs**

Image processors can be invaluable when you are manipulating graphics on the computer. While some of them have rudimentary paint program tools their main function is to modify existing graphics. The graphics can come from a paint program, a digitized image, or just about anywhere as long as they are stored on disk. You can use an image processing program to clean-up or enhance a poor quality image or distort it for a special effect. While you can use an image processing package to produce literally thousands of special effects the primary use for image processing software is to change the format, color-palette or resolution of an existing image.

**Titling Software**

A titling package is the computer version of a character generator. Most titling packages give you a number of font styles, sizes, colors, shading and special effects. They let you decide how you want the text to appear on the screen and sometimes how it will get there in the first place (dissolves, scrolls, crawls, cuts, etc.). The main function of titling software is to put words on the screen (computers are very good at putting words on screens). A titling package is probably the most valuable software a desktop video producer can own. I'll talk more about titling software in Chapter 8.

**Font Packages**

To go along with titling software there are companies that do nothing but create special fonts. Some font packages contain hundreds of different styles of text. Basically, a font package contains sets of customized alphabets that can be used by other programs when you want to put words on the screen. I'll be talking about titling and font packages in more detail in Chapter 8.

**3-D Rendering Software**

Where a paint program is a "free-hand" way to put graphics on a screen, a rendering program lets you "describe" an object mathematically and have the computer draw it for you. This is sometimes called Structured Drawing.

You can't just say "draw a tree" and have the computer do it for you but you can say "here are three points, draw lines connecting them to make a triangle" and the computer will do it. You can then say "move the triangle to the left" and the computer will do it. When you tell the computer to manipulate an object it is then rendered in it's new aspect. With a rendering program you can have the computer draw hundreds of shapes and then manipulate them for you. Objects can be moved, turned, spun around an axis, distorted, shrunk, filled in, etc. Because the computer creates the shapes mathematically it is easy for the computer to manipulate them. Unfortunately, it can take thousands of points to describe a tree shape. And even though the calculations to move a single point are relatively easy when you want to manipulate thousands of points it can take the computer a long time to redraw the object.

Structured drawing packages have many levels of realism.

- Simple lines (most CAD packages use this form).
- Wire frame (lines using a 3-D perspective).
- Filled (where all facets of a shape are filled in with color).
- Shading (where shadows and perspective shades are added to the filled image).
- Ray tracing (where reflections, surface texture and multiple light sources are all part of the calculations).

I'll talk more about rendering packages in Chapter 7.

**Ray Tracing Packages**

Ray Tracing packages are the ultimate in 3-D rendering software. Shapes are described to the computer mathematically (the software has tools to help you do this). You then describe the types of surfaces you want to use, their reflectivity, and finally where the light source(s) are in relation to the objects. When the image is rendered the computer uses a technique called ray tracing. If each ray of "light" were a straight line the computer calculates where it would strike an object, where it would bounce and what would reflect back to the observer. Once all these calculations are done the next ray of light is calculated and the next and the next. It can take the computer a long, long time to do these thousands of calculations (literally days in some cases!) but the results can be spectacular. I'll talk more about ray tracing software in Chapter 7.

**Animation Software**

There are a number of ways that a computer can animate objects. You could create an object with a paint program, turn it into a brush and then move it around the screen using your mouse.

**Color Cycling** (another paint program technique) is similar to the way that flashing lights on a Las Vegas sign simulate motion. **Page flipping** or **Cell Animation** is the "traditional" cartoon type of animation. An object is drawn on one page (or screen or frame) and saved, then the object is re-drawn in a slightly different location on the next page and saved again, etc. When you have a series of pages that are all slightly different the pages are "flipped" rapidly creating the illusion of motion. The next two types of animation could only be done with a computer. The first is called **Metamorphic** or **Transforming** animation. You first create an object using a series of points (the same way that rendering programs do) and then draw another object using the same number of points. The computer then moves the points in the first object until they match the points in the second object. In other words, you could draw a car and have the computer twist and distort it until it became an airplane. The other type of computer assisted animation is called **Tweening**. You first draw an object in one position in a frame, then in a different frame you draw the same object in a new position, the computer then fills in all the in-between frames for you. Finally, there are animation programs that combine a number of these techniques.

**ANIM files**

When you talk about animation software you will also run into something called an **ANIM** file. This is a special way of saving an animation to disk. Instead of saving each frame as a separate picture file only the first few frames are saved with all the information while the remaining frames contain only the information about what has changed in the frame. If the animation has lots of action and lots of things changing from one frame to the other the ANIM file will be very large, but if there are only one or two things moving in the scene then you can store a great deal in a smaller space. I'll be talking about animation programs in more detail in Chapter 7.

**Presentation Software**

Once you have your text and/or graphics created you will probably want to do more than just popping them onto the screen (which isn't a bad technique really). Presentation software lets you decide how you want your images to get on and off the screen. Presentation software does with computer images the same thing that an **SEG** (special effects generator) does with video images. They usually have dozens of transition effects like fades, wipes, dissolves, rolls, flips, scrolls, crawls, etc. Sometimes a titling package contains the presentation software and sometimes it is a separate module. Presentation packages can usually manipulate any type of computer graphics (text, paint program images, rendered images, digitized pictures, etc.) I'll talk a bit more about presentation packages in Chapter 8.

**Music Software**

Music programs obviously help you compose and play back music. They range from very simple tune makers (which are just about idiot-proof) to highly complex composition and sequencing programs for professional musicians. Some music programs let you connect your computer to other musical instruments through a **MIDI** device (Musical Instrument Digital Interface). Some will generate sound effects, even let you digitally record sounds with a sound sampler and turn those sounds into electronic instruments. Basically all music programs let you create and playback songs one way or another. I'll talk about music programs in Chapter 9.

**Miscellaneous Software**

There are other programs you might consider using in your desktop video world. A word processor will help you write scripts, letters, screen plays or just keep notes. A database program can be used to keep track of all your tapes, scenes, customers or equipment. A spreadsheet program can be very helpful when budgeting for a production. Tax programs are a life saver if you are running your own video business. And games are great when you just need to forget about work for a while.

# What You'll Need

So there are hundreds of pieces of software available that do hundreds of different things. What software do you really need for desktop video? Like anything it depends on what you want to do.

In the desktop video world the most popular category is titling software. If all you want to do is put titles on your videos then you should be able to get by with one or two titling programs. They are usually comprehensive, stand-alone products that will create titles for you and give you some transition options. To get a little fancier you might consider a presentation package that will give you more transitions to pick from.

A very close second to titling packages are paint programs. Even if you don't have artistic abilities a paint program can be invaluable. With a paint program you can do some titling, clean up digitized images, create special effects, even do a little animation. Paint programs can also be used for real-time presentations and are particularly useful for training and educational videos.

If you are using more than one type of computer graphics software then an image processor is probably a good idea. Not all software will tell you on the box which other packages it is compatible with. If you buy something and find out you can't share images with another of your programs an image processor can usually help. In desktop video the primary uses for image processing software are format conversions and color manipulation. Lets say you have a great graphic that you want to genlock over a video but the graphic uses all the colors in its palette. Genlocks need to have color zero free (someplace for the video signal to come through). With an image processing package you can modify the graphic to free up color zero and still have it look good.

Next on the list is probably animation software. The newer packages make it relatively easy to create and play-back animations. They do take some time to master and the results are closer to cartoons than the opening computer generated credits for Monday Night Football or Entertainment Tonight. Most of the animation packages are completely self contained which means you don't need additional software but you are locked in to their style and format. While you can usually import backgrounds from other software most of the animation packages use unique techniques for playing back the final animation. If you are thinking about animations you will probably be pleasantly surprised by what you can do with the software that is out there.

Halfway between the all-in-one animation packages and the high end graphics software you have page flipping software. With a page flipping package you can animate just about anything but with the extra flexibility they place more burden on your creative abilities.

Getting a lot fancier (and harder to master) you have the rendering and ray tracing packages. These programs, when used by a professional, can create amazing three-dimensional graphics and animations. With a ray tracing program you can create the Monday Night Football type of computer generated images. Be warned, however, that 30 seconds of that kind of animation can take months of hard work.

Music software has one very big advantage in desktop video. With a tiny bit of talent you can create background music for your videos. They may not be top 40 hits but if you create them you don't have to worry about copyright problems. Like other software you can get music software that ranges from entry level self-contained software all the way up to professional composition software.

## The Basic Package

If you are just starting out I would suggest you buy three basic packages. A titling package, a paint program and a presentation package. The fourth package is up to you. Animation and music would be next on my list then an image processing package. Finally, page flipping, rendering and ray tracing packages. Of course you may find a need to buy these in a different order depending on your application.

## Paint Programs

No matter what your video application is, you should probably buy a paint program. Fortunately, the Amiga is just about the best consumer level computer for graphics. Since the Amiga has custom graphic chips, it has a great advantage over any other home computer. Yes, you can coax an IBM or Macintosh into producing great graphics, if you have enough money, but an Amiga straight out of the box is a remarkable graphics tool. Paint programs for the Amiga are very advanced and very powerful but even a complete novice can use them without months of training. At their simplest, you can pick a color with a point and a click. From then on the mouse is your brush and the screen is your canvas. Flip through any Amiga magazine and you will see images created on the Amiga that couldn't have been done a few years ago.

Even the earliest paint programs for the Amiga were incredibly powerful and they have continued to get better and better. Competition between paint program manufacturers has meant a wide selection of high quality programs to pick from. While there are many paint programs to pick from you can be pretty sure that whatever package you buy will do just about any job you ask of it. There are, of course, a few things that you should look for in a paint program when it comes to desktop video.

It should save and load pictures in IFF format. 99% of the paint programs for the Amiga already do this but it is still a good idea to check.

There are two other considerations when using a paint program for video. You need to be able to display your graphics in interlaced and overscanned modes. While not essential except in the final display it is easier to get a feel for the final result if you can also create your graphics in these modes to begin with.

Video is an interlaced medium and some genlocks won't even function unless the Amiga graphics are interlaced too. Your Amiga monitor may flicker annoyingly in interlaced mode but you won't notice it once it is on video tape.

Video signals also produce overscanned images. Unless you want a colored border around your screen you need to be able to display graphics in overscan mode. If the border color is color zero (see Genlocks and Color Zero below) you can minimize the effects of a non-overscanned graphic by painting all the way to the border with the color zero color. Again, this is an area where an image processing package can come in handy since they can usually change the display modes of any graphic.

Extras that you will find useful (but not critical) are **anti-aliasing** and **dithering** capabilities. These are fancy names for ways to make images appear smoother and cleaner. Anti-aliasing is a very good feature to have in a titling package and is nice to have in any graphics program. Lockable palettes are handy when you are working with digitized images. You can bring in an image, lock the palette to that, and then not have to worry about unexpected color shifts when you combine the two. Some paint programs have titling features, animation features, and some will even let you do mapping (where you take an image and conform it to another shape. Like a digitized image wrapped around a sphere shape). All paint programs should give you the basic tools for generating graphics and most of them will give you a truckload of extras that you will just have to experiment with.

## Choosing A Paint Program

As I mentioned, just about any paint program for the Amiga will give you everything you might need and more but here are a few of the best paint programs.

### Deluxe Paint III from Electronic Arts                                $159.95

Electronic Arts has dominated the Amiga paint program market since it first introduced the original Deluxe Paint years ago. D-Paint (as many people call it) has gone through a number of modifications until it is now up to Version III. Deluxe Paint is also the number one best selling software for the Amiga in any category. There is a good reason for that. It is a remarkable program in many ways. The latest version offers full overscan and interlaced modes. The only mode not supported is HAM (Hold and Modify) but it does support the extra halfbrite mode (if your Amiga has the right chips). It has extensive animation capabilities using multiple screens and a form of cell animation. Of course there is the traditional paint program animation technique of color cycling. D-Paint III also offers many titling capabilities including some you may not find in stand-alone titling packages (like bending, stretching and gradient fills), There is also 3-D perspective and 3-D rotation of brushes, stenciling, some special effects and the full range of features normally found in a good paint program.

Deluxe Paint III is probably the very best paint program available for any computer and it will certainly do just about anything you need it to. Many Amiga and video people consider D-Paint an essential piece of software to own. You won't be disappointed if you buy D-Paint. It is easy to learn and very powerful.

## Photon Paint 2.0 from Microillusions $149.95

Another paint program that does support HAM mode, interlaced and overscan modes is Photon Paint. Along with all the normal features you find in a good paint program it also has a video mode where all menus and pointers will disappear. You can also instantly flip back and forth between NTSC and PAL formats, between interlaced and non-interlaced and between overscan and non-overscan. So, with only a few key-strokes you can quickly switch to video-ready graphics. Photon paint also offers a number of special features like contour mapping, shadowing, rub-through, pantograph, blend and even a form of ray tracing. There are also a handful of animation features like color cycling, page flipping and the ability to build ANIM files. Finally, there is dithering and anti-aliasing.

Photon Paint is strongest in its graphic tools and special effects. While it does have a handful of text features, including support of "colorfonts", it is not a titling program (although you could probably do more with text once it is on the screen than you could with other paint programs).

## Digi-Paint 3 from NewTek Inc. $99.95

Digi-Paint 3 is another one of the few programs that support HAM mode drawing. In HAM you can use all 4096 colors that the Amiga can produce. HAM mode will give you much more to work with but not all software packages will support HAM files (they are different than standard IFF files). You can, of course, save files in IFF if you want to. Digi-Paint 3 also offers a number of image processing features like anti-aliasing, dithering, color re-mapping and the ability to convert pictures between resolutions. You can even colorize black and white images. Beyond the usual paint program features you can do texture mapping, text rendering and transparent painting. Digi-Paint 3 also has a genlock mode that locks out color zero (so you won't accidentally use it in a painting) and it also supports interlaced and overscanned modes.

Some of the tools are a little complicated or awkward to use (compared to other paint programs) and Digi-Paint 3 doesn't offer much in the way of animation but Digi-Paint 3 does double duty by giving you many more advanced image processing features.

## Image Processors

Here is where things get a little fuzzy. Some paint programs have image processing features while some image processors, animation packages and even presentation packages have paint program features. Rather than try to draw a definitive line between these different packages I'll just go ahead and list a few of the best.

### Deluxe PhotoLab from Electronic Arts $150.00

This is a hybrid package that includes a paint program, poster printing program and image processing program. The paint program is complete but you won't find quite as many features as you would in a dedicated paint package. It does, however, support HAM mode, which Deluxe Paint does not. There are almost no text or animation features. The poster printing section of Deluxe PhotoLab will let you make almost any size poster from almost any graphic using almost any printer, but as far as desktop video goes you probably won't ever use this feature. The image processing portion of the program is very powerful and will be the portion of the package most used by desktop video people. The nicest feature of the image processing program is the statistics display. Here you can see and manipulate the picture down to the pixel level. You can blend colors, average colors, free-up color registers, change picture size and resolution, and more.

It might be a good place to start, if you know that you aren't going to do a lot of fancy paint operations. You could get by with Deluxe PhotoLab as your only paint and image processing software. As a compliment to any of the other paint programs Deluxe PhotoLab is a stand out program.

### Butcher from Eagle Tree Software $37.00

Butcher is an image processing, conversion program plain and simple. It supports and can convert to and from all resolutions and modes. There are a number of useful features for video work such as the ability to automatically free-up color zero with a single command (if there are pixels using color zero they will be changed to the next nearest color). There are a few special effects, such as turning an image into a mosaic pattern, false colors, toning, edge mapping, even a command to turn a color image into a simulated "old fashioned" sepia tone photograph. Butcher's biggest advantage is the number of essential features you get at the price. If you already own a paint program and are looking for an inexpensive image processing program, Butcher should be able to handle the job.

### PIXmate from Progressive Peripherals and Software $70.00

Like any good image processor PIXmate supports and can convert to and from all resolutions and modes, offers just about any color and pixel manipulation task you might need, plus it offers over 3,000 special effects (although most of them are variations on the "normal" color manipulation operations). There are not a lot of "extras" in PIXmate but there are hundreds of normal image processing options making it the most powerful program of this type. The odds are, if you need to do any kind of enhancement, modification or manipulation to an image, PIXmate can do it. It is a very powerful program and will probably take most people a fair amount of experimentation to get the maximum use from it.

It would be easy to carry on with descriptions of products that have built-in painting features or image processing abilities but we have to draw the line somewhere. All of these products are considered the best of their type and all are worth owning. Each of them have their own special advantages and disadvantages. All of the paint programs are fairly easy to learn and use even though they are powerful tools. The image processing programs are a bit more complex at first but all have fairly good manuals and you should have no trouble using the basic features.

## Paint and Image Processing Program Tips

When using a paint or image processing program to produce images for later use in video there are a few things that you have to keep in mind from the very beginning. I'll talk about a few of the most important of them here.

**Adjust Your Monitor**  No matter how great your graphics look, if your monitor is not properly adjusted the final result will not be what you were expecting. The best way to adjust a monitor is to start with a color bar generator (outputting pure RGB).

Next, there are programs which will generate color bars for you through software. You can sometimes record color bars off the air (usually just before or just after a station is ready to go off the air), you can then play back the tape through the monitor. Next, you can buy color bar charts, set one in front of a properly adjusted camera and feed the video signal into the monitor (obviously this method has it's drawbacks because you are assuming that the signal reaching the monitor is correct). Finally, you can feed a good video signal through the monitor and adjust it by "feel."

There are two methods for adjusting a monitor or TV. Before starting, you should let the monitor warm up for a few minutes before making adjustments because the colors will drift slightly when the monitor is first turned on. The first method is to start with the best possible signal feeding into the monitor, turn down the color and tint (or hue) controls until you have a black and white picture, adjust the brightness and contrast until you have nice white whites, black blacks and smooth grays in between. Next, turn up the color control until the saturation looks normal. Finally, adjust the tint control until the greenish tint turns to a redish tint. The second method is to send a clean signal to the monitor that has a lot of flesh tones (Caucasian faces work best). You can usually get good flesh tones from newscasts or soap operas. Adjust the brightness and contrast as described above, then turn the color control up too high. Now adjust the tint until the flesh tones look right, and finally, bring the color back down until the overall picture looks good.

**Read and Experiment**

You will find that there are hundreds of tricks and techniques for producing different results with a paint program. Each paint program offers a number of features to make the job of creating images easier. The most important tip on using any paint package is to read through the manual and try out each of the options. The manual may be well written but until you actually try a feature you won't know what it can do for you. These extra features can help just about anyone create impressive graphics, no matter how good or bad an artist you are.

**Know the Limits**

Computer graphics have come a long way in the past few years but they are still far from being able to fool the human eye. Unless you are an exceptionally good computer artist people will be able tell a computer generated image from a camera generated image instantly.

This doesn't mean that computer graphics are bad for graphics, it just means that realism is not the computer's strong suit. Computers are, however, very good at doing graphs, charts, diagrams, geometric shapes and patterns. They are also good at manipulating graphics and colors in ways that would be impossible with a brush and canvas. They are not very good at doing smooth curves but great at doing straight lines (horizontal and vertical ones that is). If you need realism then use video or digitize an image. Remember that a computer generated image is going to look like a computer generated image no matter what you do. Finally, paint and image processing programs are great for producing special effects, just don't get carried away putting special effects in every other scene. A particular effect may look great but if it doesn't fit with the rest of the production then don't use it. Special effects aren't very special if you over-use them.

**Keep it Simple**

Even though a video picture can be very complicated and detailed you will find that computer graphics are more effective if they are kept fairly simple. Charts and graphs are best if you limit the total information content of a screen to two or three elements. If you need to convey more than that then break your information into more screens. Spend a few hours watching educational TV and you will find that all the graphics are usually pretty simple. You don't have to make everything understandable by a 5 year old but don't bombard the viewer with too much information all at once.

**Genlocks and Color Zero**

When you start mixing graphics and genlocks you will need to be able to specify (or at least know about) color zero. Genlocks consider color zero to be "invisible". Any shape painted with color zero will let the video come through as if you had cut out that color with a pair of scissors. Some genlocks will let you change which color is considered "invisible" but most use color zero. Note: color zero is an Amiga term NOT a specific color. In other words, color zero could be black, red, blue, chartreuse, or any one of the 4096 colors available on the Amiga. It is better to think of color zero as a special position in your palette rather than a color.

All paint programs use the color zero register in one way or another but not all of them will tell you in the manual which one it is. When in doubt, it is probably the upper left or far left color in the palette. You can always find out which is color zero by trial and error if you have to. Just paint a number of colored bars and try to genlock the image with different settings on the genlock. The bar where the video comes through first is the color zero position on the paint program palette.

If you are genlocking graphics created with another program you may have to use an image processing package to rearrange the palette and free up color zero.

**Illegal Colors**

Perhaps one of the most common problems with integrating the Amiga with video is one of the least understood. Of the 4096 colors that the Amiga can reproduce not all of them are suitable for video. The Amiga is quite capable of producing illegal colors (illegal as far as NTSC standards are concerned).

Normally, when you analyze a video signal on a waveform monitor the intensity of the picture should never exceed 100 IRE (Institute of Radio Engineers). On a waveform monitor (that is properly calibrated) sync will be at -40 IRE on the scale, black at 7.5 IRE and 100% white should be at 100 on the scale. However, the Amiga can produce colors that climb far above 100 IRE on the scale. These colors look fine on an

Amiga monitor but when these colors are put on video tape the results can mean a terrible picture and they can even effect the audio.

The problem is further compounded by a number of factors. First, there are a number of poor genlocks and encoders for the Amiga that just make the problem worse. Second, you have to look at the total white level. It is possible to use perfectly legal colors from the Amiga but when you use them together or add them to another video signal the total climbs above 100 IRE. Third, some combinations of legal colors when placed next to each other on the screen will produce values over 100 IRE.

Unfortunately, unless you have a waveform monitor, there is no simple way to avoid these problems. There are a few ways to minimize them though. The first step is to start with a good genlock (see Chapter 3). The next thing that you can do is limit saturation or intensity levels of the colors that you use. Just about all paint programs let you create your own palettes and modify each color. Usually, you modify the colors using slider bars to adjust the RGB values, some programs will also let you adjust the hue, saturation, contrast, intensity or value (many programs use different names for these controls). A pure white color would have all the sliders at the top of the scale. Each program uses a different range for these scales, some go from zero to 10, while others go up to 16. The best rule of thumb is to keep your color values below 75% or 80% (particularly the whites and reds).

Try to stick with greys instead of whites. Pastel colors work better than primary or fully saturated colors. Another trick is to create a tight pattern of complimentary colors to fill an area rather than use a single, solid color. But you should avoid placing contrasting colors side by side (this is where anti-aliasing and dithering helps). We have all seen people on television wearing herring bone or shark skin suits that turn "electric" on camera. Image processors are great for adjusting colors after the fact, and some of them will even let you adjust the overall intensity of a screen. Finally, you can try genlocking your graphic on top of a video image without audio. Turn up the volume louder than normal before you bring up the computer graphics. If you notice a significant increase in the hum or you hear a loud buzz as the graphics are added then your colors are probably oversaturated.

**HAM and Image processors**

Image processors will work with any image but HAM mode graphics are a special case. Because of the tricks used to generate a HAM image you can't be sure what will happen when you try to adjust the colors. It is better (and faster) to convert a HAM image to 32 colors (or 64 colors if you have extra halfbrite) before manipulating the image. All image processor manuals have a section about the different graphic modes. It is a good idea to read through it once (even if it doesn't make much sense).

Obviously, I haven't covered even a small fraction of the things that can be done with a paint program or image processing package. Entire books have been written about paint programs and computer graphics. Magazines include articles about computer graphics in just about every issue. The thing to remember about paint programs and computer graphics is that to get the most from any tool you should practice and experiment. Next to a titling package, a paint program is the most valuable and versatile piece of software a desktop video producer can own.

# Chapter 7

## Rendering Ray Tracing and Animation software

# Chapter 7
# Rendering, Ray Tracing and Animation Software

Everyone who knows about graphic packages has their own ideas about how they should be divided into categories. I could have lumped them all into one large chapter covering paint programs, titling programs, animation, rendering, image processing, presentation and ray tracing programs. Or I could have given each one a chapter of their own. In one way or another all of these packages are designed to do different things with graphics. In this chapter I'll talk about Rendering and Ray Tracing packages first and then get into Animation software.

## Structured Drawing Rendering and Ray Tracing Programs

Beyond paint programs the next step up in graphics software for desktop video are rendering packages. Rendering software can be considered structured drawing software. Structured drawing can be as simple as line drawings or as complicated as ray tracing. The main difference between a paint program and a structured drawing program is that in a paint program the graphics are temporarily stored on the screen, where a structured drawing program stores the graphics point by point in memory. In other words, as you create an image with a paint program each piece is added to the screen as you draw it. The program only remembers one or two steps at a time as you do this (which is why UNDO only works on the last thing that you did). Once the pieces are added to the screen they become part of the total graphic. When you go to save the image to disk the entire screen is saved as one unit, not as a collection of parts. A structured drawing program works in a different way. As you create parts of a drawing they are defined mathematically and stored in a list in memory, they are then drawn on the screen for you. Each part is stored as a separate unit. When you add a new piece to a structured drawing it is added to the list and then the entire screen is re-drawn. When you go to save the image to disk the mathematical descriptions are saved in a long list (although most structured drawing programs will also let you save an image as an IFF file if you wish). When you load the image back into the program the computer has to re-draw the entire screen piece by piece.

The advantages to a structured drawing program are that each piece of the image can be manipulated separately. Since the objects are "described" as a series of points the computer can manipulate objects in hundreds of ways. Not only can you move objects up, down, left and right on the screen you can enlarge them, shrink them, turn them, spin them, twist them, give them perspective and you can do all this in three dimensions. You can also do things with a structured drawing that you just can't do with a paint program image. With a ray tracing program (the ultimate in structured drawing programs) you can have the computer create very realistic images automatically. Another advantage is in creating animations. Since each part of an image can be manipulated separately and automatically with a great deal of precision you can tell the program what, where, when, how and by how much to change something in an image.

The disadvantages of structured drawing programs are that as the image gets more and more complicated it takes the computer more and more time to manipulate the parts. Another disadvantage is that since the objects are stored as a series of points with connecting lines it can take a great many points to describe a curved or irregularly shaped object. Finally, structured drawing programs are harder to master than paint programs. The fancier they are the harder they are to learn.

## Choosing a Rendering Program

When talking about rendering programs in a desktop video environment there are a few things you should consider. Interface, speed, options, formats and animation capabilities. In some cases you might also want to look at how the program deals with text. All rendering packages will let you manipulate text if it is created in the package but a few will give you special text manipulation features.

There is no easy way to divide these programs into categories. All ray tracing programs are rendering programs but not all rendering programs will do ray tracing. Most rendering programs will give you some animation capabilities and some animation programs can do rendering. Some so-called animation programs are really rendering programs and some rendering programs concentrate more on animation features. There are also some presentation programs that have paint and rendering features and most of them have some sort of animation capabilities. I hope that clears things up. Anyway, here are a few suggestions.

## Sculpt-Animate 4D and Sculpt-Animate 4D Jr. from Byte by Byte
## $499.95 and $149.95

You can't talk about rendering software without talking about Sculpt 4D. No question that Sculpt is at the top of the list. Sculpt-Animate 4D is probably the most powerful rendering package ever created. It supports all graphic modes plus dithering and anti-aliasing. It offers just about every feature you can think of plus a few hundred more. Creation of objects is done in three windows, each giving a separate view. There are tools that automate just about every step in creating an image. You can twist, turn, distort, mirror, copy, move, extrude and alter objects in dozens of ways. Input can be "free-hand" with the mouse or you can enter individual points with exact coordinates. When it comes to rendering, Sculpt offers another slew of options from simple wire-frame all the way up to phenomenally impressive ray tracing. You can specify the colors, light sources and reflective qualities of an object. You can simulate metal, glass, plastic or any number of surfaces.

Sculpt-Animate 4D also includes a number of advanced animation features. For example you can specify the tracks and paths of individual objects, you can also specify "camera" movements (as if the object was stationary and a camera were moving around it). When all the tracks are set you just sit back and let the program do the rest. It automatically, renders each frame and stores it in an ANIM file then it does the next frame and the next. You can also render to a frame buffer or with the right controller send each frame directly to a VCR.

While Sculpt 4D gives you many tools that simplify the steps of creating, rendering and animating be warned that it is not a quick or simple process. You are going to have to spend a fair amount of time learning how to use the program. The manual is very well written and with time and experimentation you should be able to create some spectacular results. Also, while Sculpt 4D is one of the fastest rendering packages available, the more complicated the image the longer it will take (some images can literally take days to finish). Serious animators usually end up purchasing accelerator boards and as much extra memory as they can afford. If price is a problem Sculpt-Animate 4D Jr. does just about everything that Sculpt-Animate 4D does except ray tracing. Of course, ray tracing is the ultimate in rendering techniques and you will probably end up upgrading if you get serious. If you think of a paint program as a rowboat then Sculpt-Animate 4D is a nuclear battleship. Much more complicated but much more powerful.

### Turbo Silver 3.0 from Impulse $199.00

Turbo Silver is about the only rendering program that can compete with Sculpt as far as features and the final results. With dozens of tools, features, modes and extras, Turbo Silver is a very powerful program. Creating objects is done in three screens each showing one perspective view, but only one screen at a time is visible. Silver works in all modes and resolutions, has numerous tools for helping create objects and dozens of ways to render the final image (this is the programs main strength). Silver has features that let you create animations automatically. It also has a texture mapping feature that lets you wrap an IFF brush to the surface of an object.

Silver, like any powerful rendering program, requires a fair amount of time to learn and the manual could be better. There are times when many steps are required to perform what would seem to be an easy task. This is more due to the fact that there are dozens of options available with almost every operation, rather than a design mistake, but it can be frustrating. Expect to spend hours creating and rendering objects in the beginning and, naturally, with more complicated objects the rendering times can be very long.

There are people who swear that Silver is the best rendering program while Sculpt has the best interface and creation tools. There are others who claim the reverse. Either way, impressive results can be achieved with either program and a fair amount of time and persistence.

### Page Render 3D from Mindware International $159.95

Page Render 3D is unique in that it lets you create your objects in a single window with a three dimensional perspective. This feature can make it easier to design 3D objects quickly. The program supports all modes and resolutions, offers numerous tools for creation and rendering (only a pseudo-ray tracing is supported but the results are quite good) and a unique scripting language enhances animation possibilities.

Page Render 3D may not be quite as high-powered as Sculpt or Silver in some ways but it offers a completely different set of tools that will let you do things quickly and easily. There are enough built-in effects (such as twisting an array of objects into domes, funnels, spheres, even sine waves) that you might never need to create your own. For 3D text manipulation Page Render is very effective.

While there are some very powerful features in Page Render 3D (particularly the arrays, scripting language and animation features) its biggest advantage is that it is one of the easier rendering programs to learn. However, you should still expect to spend a fair amount of time learning the program.

## Animator:Apprentice from Hash Enterprises $299.95

One of the first 3D rendering/animation programs for the Amiga is Animator:Apprentice. Compared to some of the other programs it is a little rough around the edges but it offers some features that you won't find anywhere else. One of the highlights of Apprentice is the way that it animates rendered objects. You start with a frame or skeleton, determine the actions and the surfaces that will go on the skeleton. Then the program maps the parts onto the skeleton for each frame. The results are not perfect but you can go back and clean things up after the fact. Apprentice is one of the few rendering package that does accurate mapping using topological mapping (rather than just projecting a brush onto a shape and thereby introducing distortion). There is an updated version of Apprentice, that should be out by the time you read this, that uses a different form of rendering, (something called spline patches), which uses curved shapes rather than a multitude of triangles.

There are a number of other rendering programs out there. Some of them offer features that you can't find in the programs mentioned. Some are faster, easier, harder or just different. Modeler 3D from Aegis, Forms in Flight II from Centaur Software and 3-Demon from Mimetics are all good programs that many people use for video work (although they aren't quite as sophisticated). If you plan to get heavily involved with rendering software then you will probably need extra memory (at least 1 megabyte and probably 3 megabytes or more) and a faster processor and math chip (68020 or 68030 and a 68881). If you end up with more than one package you will find that each package uses a unique format for saving objects. There is a program called InterChange from Syndesis ($49.95) that will convert objects from one format to another. Interchange is like an image processing program for rendering packages. It doesn't do any modification to the objects but will let you create objects with one package and render with another.

Any of these packages will let you create impressive graphics once you learn how to use them. The objects can be very realistic but will have a "science-fiction" or "hi-tech" look to them. Some types of objects, (particularly things that are mirrored, glass, machine-like or mechanical) work well. Rendering packages are good at producing spheres, cones, boxes, cylinders, buildings, geodesic domes, patterns, mathematical shapes, etc. Rendering packages are also very good for 3-D titling animations. However, things like faces, natural objects or irregularly shaped things will still come out looking mechanical.

The results can be impressive but once more I will mention that they will take a fair amount of time. Time to learn the program, time to create the objects, time to render them and time to animate them. I'm not trying to scare you away from rendering software but they are all very complicated programs even for the computer wizard, professional graphic artist or animator. On the other hand, if you can spend the time, effort and money to master these programs then you have a very special, very powerful video tool that other people will pay money for. It may take weeks or months to create thirty seconds of ray traced animation but companies will spend thousands and thousands of dollars for those few seconds. There are companies that do nothing but rendering style animations on multi-million dollar super-computers. Right now, the Amiga is one of the few computers that can do these same types of graphics.

## Animation Software

There are many animation packages available for the Amiga. There are also many programs that have animation capabilities built in. There are also presentation programs that feature special types of animation. I've already talked about some programs with animation features but in all of them animation is secondary to their main function. This section will deal with programs whose main function is to help you create animations with the Amiga.

## Types of Animation

In Chapter 6 I briefly described some of the different types of animation that can be done with a computer. You could say that computer animation is anything that moves on the screen. Let's take a look at the different types of computer animation in more detail.

## Hand Controlled Animation

You could say that the simplest form of computer animation isn't really animation at all. It is closer to puppetry than animation but it is easy to do and can be effective. In this form of animation you have a static screen and the only thing that is moving is the mouse pointer. Let's say you are doing an educational video about a car engine. You could put a picture of the engine on the screen (digitized, drawn or even a video image that is genlocked). Then, as the narrator talks about different parts of the engine you move the pointer around the screen pointing at the various parts. Getting a little fancier, you could customize the pointer either using the Workbench Preferences (see your AmigaDOS Manual on how to do this) or by creating a custom brush with a paint program.

A variation of this type of animation is to draw while you are recording. One of the oldest movie tricks used to show a long journey was to show a map or globe and have a little airplane flying or line being drawn from the start point to the destination. Weather people on just about every news program use this simple type of animation to draw in clouds, fronts, rain, etc.

One of the most fascinating things to watch is an artist drawing a picture, even more fascinating is watching a computer drawing a picture. If you are a good artist this can be very effective. If you have a structured drawing package you can take your time creating a very detailed drawing and then record as the computer redraws it. Another trick used in early TV was to paint with the chroma key color letting another video image come through. You can do this with a genlock and a paint program by filling the screen with one color, genlocking the Amiga with an external image and then "painting" with color zero. Where you paint, the external image comes through.

Just because these don't sound like "real" animation techniques don't dismiss them altogether. If all you need is a few seconds of something moving, pointing, drawing a quick line or a simple shape, it might take hours to set this up with an animation package but only take a few minutes to do it by hand.

## Color Cycling

One of the earliest forms of animation done on the Amiga was with a technique called color cycling. Color cycling is a computer trick that is similar to the way motion is simulated on a marquee or "moving" neon sign. By rapidly turning lights on and off in a particular pattern the eye is fooled into seeing motion. Color cycling takes this technique one step further. Instead of rapidly turning pixels on and off (which you can do if you like) the computer rapidly changes pixel colors. Lets say you draw a line using four colors 1,2,3 and 4. You then rapidly shift the colors so that color 1 changes to color 2, color 2 to 3, 3 to 4 and 4 back to 1. Even though nothing is actually moving on the screen your eye is "carried" along the line following one color as it "moves." The simplest form of color cycling is, as mentioned above, cycling between two colors. This can create a back and forth movement. The easiest way to create color cycling is to paint in the cycling mode (if your paint program allows this). Otherwise you will have to change your palette manually. While color cycling can be used to do very short animations (the number of positions are determined by the size of your palette) the real strength of color cycling is in small repetitive movements like you find in running water, crackling fires, etc. You can also create dizzying op-art effects with geometric shapes or patterns that are color cycled. There are also titling packages that will cycle colors to create a rippling effect as if a light source was moving across the text.

**Page Flipping**

The very first "moving pictures" used page flipping techniques to simulate motion. A series of pages, each with a slightly different still image, were flipped in rapid succession. Just about all cartoons are done using this type of animation and even films and video could be considered page flipping.

Computer page flipping programs do the same thing with screens. Each screen has a slightly different picture and when rapidly changed the illusion of motion is achieved. Smooth motion takes about 12 to 15 frames per second otherwise the eye detects the flicker. These pages or screens can be created in any number of ways, with a paint program, rendering program, digitizer, or just about anything that creates an IFF image. With page flipping programs it is a good news, bad news situation. The good news is that if you can get the graphics into the computer a page flipping program will animate them for you. That means that you could animate almost anything from any program. The bad news is that you have to figure out how to get all those slightly different images created and saved somehow. Most page flipping programs offer features to help you put the animation together, arrange sequences, adjust the speed of the animation, synchronize digitized sounds or music, and sometimes they have compression routines (to reduce the size of the animation files when they are stored on disk), but they don't really help create the individual pages for you.

**Cell Animation**

A slight variation on the page flipping technique is cell animation. In traditional animation the cells were transparent sheets (cellulose film) that could be overlaid on a background illustration. The object to be animated would be drawn on a series of transparent cells. A cell would be placed over the background illustration and that page would be shot with a movie camera. The next cell would replace the first one (on top of the same background) and shot. This way the entire scene wouldn't have to be redrawn for each shot. To have a cartoon character walk across a scene the artists would first create the background illustration. They would then create a series of cells with the character walking.

They would then overlay each cell and shoot each one. When the whole scene was played back the character would walk across the background. The other variation of this is to have the character cells create the illusion of walking but remain in the center of the scene. A background illustration would be created that was longer (or taller) than the screen. After each shot the background would be moved slightly. The effect created was as if the camera were following the character as it walked along. Sometimes, to save time the background illustration was looped around so that the characters would keep walking (or running, driving, bouncing, or whatever) past the same background.

Many computer animation programs use this trick. First you create a series of cells to animate the main object (ten cells might contain the movements of a character walking for example). In the computer these cells are more like brushes or stamps that you can place anywhere on the screen. You then create and bring in a background illustration. Next you place cell number 1 where you want it and click the mouse button. The program then automatically changes your brush to cell number 2 which you position and click into place. Every time you click the mouse button the computer stores that page and advances to the next cell for you. When you reach the end of the series it loops back to the first cell again. When you are finished the program can play back the entire scene for you. This way you only have to create a series of ten cells and one background, but with them you can generate many seconds of animation. The disadvantage to these types of animation programs (and others types too) is that rather than storing each frame of an animation, some programs only store the background, the cells, the timing and the path. This can be a great space saver but it also means that you can only play back the animation using that particular program.

## Tweening Animation

Unique to computers is the tweening technique of animation. The idea is a simple one but can get very complex when combined with other animation techniques. Essentially, you create an object on the screen in one place (the first position), and then copy the object to another place (the final position). You then tell the computer to move the object from the first position to the final position, automatically drawing each of the "in be-TWEEN" screens for you. The fewer tweens you specify the faster the animation. Tweening can be achieved with either a page flipping technique (where each tween is a separate page) or by drawing an object, erasing it, re-drawing it in a new position, erasing it, etc.

Tweening works well when the object doesn't change as it moves (like an airplane moving across the sky, a ball falling or text moving around the screen). When you want the object to change as it moves you have to use additional types of animation. Usually, when you specify the two positions the computer just moves the object in a straight line between them. But some advanced tweening programs let you specify the path that the object will take as it moves from start to finish. Some tweening programs will also help you calculate how fast an object falls, bounces or arches with built-in trigonometric functions. Although many animation programs use tweening, they usually combine other animation techniques to create more realistic movements.

**Metamorphic Animation**

Another computer specific type of animation is metamorphic animation. In a structured drawing program all objects are composed of a set of points with lines connecting them. The points are stored in the computer's memory as coordinates. Some metamorphic animation programs will let you create an object in one position (using one set of points) and then create a second object (usually with the same number of points) in a different position or arrangement. You can then have the computer move the points of the first object until they match up with the points of the second object. You can tell the computer how many times you want the image redrawn as it moves the points (how many tweens). This way you can have a car turn into a boat, Dr. Jekyll will turn into Mr. Hyde, a bird flaps its wings as it flies or a ball will turn as it falls.

One of the drawbacks to metamorphic animation is that in order for the computer to manipulate an object it has to be in a structured format to begin with. That means that you can't just create an object with a paint program or digitizer and then have it transform into something else with a metamorphic animation package.

**Brush Animation**

One more computer specific animation technique has been called brush animation. Most paint programs allow you to outline an area and turn it into a brush. The computer basically makes a two dimensional copy of the area and lets you move it around the screen using the mouse. But paint programs also go to the next step by letting you manipulate that brush. You can turn it around an axis, flip it side for side, top for bottom, stretch it, shrink it, distort it, etc. Some programs will even let you map the brush onto another shape or mathematically extrude the two dimensional brush into a three dimensional shape. You can take these brush manipulation techniques and put them into an animation program. Instead of just changing the brush in one step the program does it in a series of steps, saving each one.

This might have limited use in traditional animation (except when the character gets squashed flat by a falling safe and drifts on the wind like a leaf) but this can be a very effective special effect when it comes to animating text.

**Screen Manipulation**

A special category of animation software is primarily used for special effects and in presentation software. Screen manipulation programs don't really do anything with the objects on the screen, pages or tweens. Instead they manipulate the computer display mathematically at the pixel level. For example, if you reduce the intensity of all the pixels at once the image fades to black. If you slide all the pixels in one direction or another the image slides across the screen. Using image processing techniques, some Amiga graphics tricks and the way the Amiga hardware manipulates screen images these programs can create some interesting transitional effects. This is the closest thing to ADO (Ampex Digital Optical) video effects you will find in consumer level video.

**Hybrid Animation**

Most animation packages use a combination of these techniques to achieve their results. They may use tweening and metamorphic animation to generate the pages and then a page flipper to create the movement. They may have a structured drawing section to create the main objects and figures but let you import IFF pictures for backgrounds. Some will only let you define a single path for an object while others will let you have multiple paths running at different times. A few of the animation programs combine cell techniques with paths during the creation of the scenes.

## Choosing an Animation Program

There are a number of good animation programs available for the Amiga. Each have their strengths and weaknesses. The three main kinds of animation software are the all-in-one programs, the page flipping programs and the presentation programs. Presentation programs will be covered in Chapter 8.

## Page Flipping Programs

Page flipping programs are the most versatile in that they can animate almost anything but all the image creation is handled by you. They are very powerful, not too difficult to learn and give you a wide range of features.

**Animation:Flipper from Hash Enterprises** $59.95
**Animation:Editor** $59.95

These two programs together will give you a set of professional tools to work with animations, either created with other Hash products or almost any other program. They could almost be considered animation processors.

Editor lets you combine and edit animations from different sources, convert resolutions and modes (including HAM, overscan, and even PAL formats), manipulate palettes, even insert IFF pictures into existing animations. Flipper works with any resolution IFF pictures, including HAM, overscan and PAL images. It lets you preview your animations or 'pencil test' them for smooth action before they are saved. It creates ANIM or custom Hash formats. It will also let you specify playback rate, frame duration, mouse control and single-stepping.

**PageFlipper Plus F/X from Mindware International** $159.95

PageFlipper has gone through a number of iterations and they keep adding features to it. It accommodates just about all modes and formats including the ability to run animations of different resolutions. There are a number of transition effects or you can create your own. It offers a selection of compaction options and you can chain animations from different disks. There are a number of features for "touching-up" animations, adding backgrounds, even combining more than one animation. You can even use PageFlipper to control sounds and music. In a way, PageFlipper is an image processor for animations in that it can change or merge animations of different formats.

**Photon Video: Cel Animator from Microillusions** $149.95

Cel Animator includes almost all of the features you would expect in a page flipping program plus it has a few extras. There are paint program tools for either creating or touching up animations, it supports just about every mode and resolution, you can specify speed, duration and sequence of the individual frames, and it has features for coordinating digitized music or sounds. One of the nice features of Cel Animator is that each frame can be displayed with frame numbers, drawing numbers and sound reading abbreviations for quick reference. You can also extract individual IFF frames from an ANIM file. Finally, you can link your animation to the Microillusions Transport Controller for single frame video recorders.

## All-in-One Animation Programs

Like most things in life, All-in-one programs have their pluses and minuses. On the negative side most of the all-in-one programs use their own formats which means that you can't move animations from one program to another (although a few will create ANIM files). This isn't too much of a problem unless the program does not support overscan or interlaced modes. It can be difficult or even impossible to convert an animation with an image processing program. Another drawback is that most animation programs only work in low resolution. You are also locked into the system that they use. On the positive side, most of the animation packages are self contained. Everything you need to create and play back an animation is included with the software (although most let you bring in graphics and music from other programs for backgrounds). The programs are also fairly easy to learn and use. Finally, most of them perform very well.

### MovieSetter from Gold Disk $99.95

This is one of the easiest all-in-one animation programs on the market. With storyboards, stereo sound support (with some included sound effects), IFF compatibility, multiple movement tracks for multiple objects, color cycling, background wipes, background scrolling and a host of other features MovieSetter is a very versatile animation program. While it does support overscan you are limited to low resolution screens. The technique that MovieSetter uses is a combination of cell animation, tweening and page flipping. A character's movements are stored in a series of cells, as you click the mouse button the cell is placed on the screen and the next cell is loaded, etc. When all of the positions have been placed on the screen you can play back the animation at any speed. You can then load in backgrounds, add other characters, define priority levels (which character moves behind or in front of what things) and then save the animation (unfortunately not in ANIM format). Another nice feature of MovieSetter are the line and elliptical guides with trigonometric calculations for realistic horizontal or vertical motion.

### Zoetrope from Antic Publishing Inc. $139.95

This program has a number of features to do everything from creating the images with a built-in paint program to animation. It supports most of the other animation formats (including ANIM files) but it is limited to low resolution images. The animation technique is a combination of tweening, brush manipulation and metamorphic. There are also a number of special pixel, color and transition effects built in. Scripts are easily created and modified letting you create very smooth animations.

**Fantavision from Broderbund** $59.95

One of the only all-in-one animation packages that supports almost all formats and resolutions, Fantavision uses metamorphic and tweening techniques to create animation. You create and set an object on the screen in its starting position then modify and set the object in its final position. The program then moves and manipulates the object as it moves along the path. The program will let you import IFF brushes (but not metamorphize them) and graphics from other programs for backgrounds. There are also features for incorporating sound effects.

## Rendering Package Tips

Since all of these programs have different features, use different terms and have different ways of doing things you will obviously have to read through the manuals carefully and experiment. About the only tips that I can give are to experiment with simple things first. You should also know that some genlocks change the clock speed of the Amiga and most animation software counts on this internal clock for timing. You may spend hours getting your animation to run at the right speed only to find that when you try to put it on video tape with a genlock it is running too slow. Try something simple, all the way from the start to the finished tape.

In a rendering package try to keep the objects as simple as possible while you are learning the program. It is very frustrating to wait a few hours while the program renders a complex image only to find that you hadn't set your light sources correctly. You should also be comfortable with static rendering before you try to use the animation features.

In the all-in-one programs try a simple ball moving across a blank screen, then spin the ball, then spin and move the ball, then bounce the ball, then bring in a background, etc. Build up slowly rather than trying to create Fantasia the first time you sit down with a program.

The page flipping programs might be considered half-way programs between the simple all-in-one animation programs and the heavyweight rendering packages. They are a little harder to learn and use than the all-in-ones but not as hard as the rendering packages.

Finally, when you get your animation finished and you are ready to transfer it to video tape, remember that it is much easier (and less expensive) to transfer one sequence at a time and edit them together later. You don't have to have twenty minutes of animation transferred all at once. Besides, you probably couldn't fit it all in the computer or on a hard disk. And just because you can't afford a single frame animation VCR you might consider the old fashioned way of doing things. Buy a super 8 movie camera and shoot your single frames off the computer screen. You can later transfer the film to video tape.

## Summary

This chapter could easily be stretched into an entire book and if you are interested in learning about computer animation you should look through the Amiga magazines. There are also quite a few books available about traditional animation techniques that will be quite helpful.

# Chapter 8

## Video Titling

# Chapter 8
# Video Titling

One of the most common and important rolls that the computer can play in desktop video is in titling. In traditional video studios they use a special piece of equipment called a character generator to put words on the screen. These devices are fairly expensive and rather limited. Usually they only have one or two fonts with one or two sizes and yet they can cost thousands of dollars. In their favor they do a very good job putting that text on a screen (better than a home computer can in most cases). Essentially, a character generator is a keyboard and a kind of computer with some memory. As I mentioned earlier computers are very good at putting words on screens and the Amiga (with the right software) is one of the better computers for doing this.

When it comes to titling software there are three variations. First, there are programs that just help you generate text and screens and save them. You could call them text/screen only programs. Second, there are programs that let you manipulate text and screens and determine how you want them displayed. These could be called display programs but are sometimes called presentation packages. Third, there are programs that perform both of these functions by helping you create text screens and then let you display them. These could be called text/screen/display programs. Each of these have pluses and minuses.

There are two other types of titling software that don't really do anything until you bring them into another titling package. These are the font packages and clip-art disks. Many titling programs give you a set or sets of fonts to work with right from the start but when you want to get fancy there are programs that contain nothing but specialized sets of fonts created by other people. There are also diskettes full of nothing but backgrounds and graphics created by artists that you can use with paint programs, titling packages, page flipping programs, etc.

**Text/Screen Generating Programs**

There are a few titling programs out there that do nothing but help you create screens for titling purposes. They don't give you any special ways to display those screens (they are either on or off and usually you can't link more than one screen together). All they do is give you the tools to create, modify and save screens. You could say that paint and rendering programs fall into this category but they usually have only limited text capabilities. Text/Screen generating programs sometimes have limited paint and rendering features but these are mainly geared toward text manipulation. However, since nearly every feature is designed to produce title screens they do it very, very well. If you have a need to create a custom title screen, use a special font in a different way or just want to do extra special screens, then these are the types of programs you will need.

A Text/Screen generating program has either none or very limited display capabilities because they are for designing and creating single screens at a time. They do, however give you much more flexibility when creating those screens. You could, of course, design a separate screen for each title, record them and then cut from one to the other. Most people will create a set of screens and then use a presentation package or page flipping program to put them all together before transferring to tape.

**Presentation (Display Only) Programs**

These programs will let you put together various parts and pieces of graphics, screens, text, even digitized music into a complete, smoothly running whole. They usually let you create some sort of script that determines the order that things will happen on the screen. You can specify what background pictures to use (and where they are located on your diskettes) what objects (or brushes) you want to use, how you want images to get onto the screen, where they move on the screen, how long they remain on the screen and how they get off the screen.

Sometimes these programs will create an ANIM file of the images and sometimes they use their own proprietary routines. They usually give you a variety of transition effects, timing, movement and object positioning controls. Presentation programs are not just for video applications so they usually include features to let a script repeat over and over again for unattended demos plus keyboard or mouse interaction routines. Presentation programs don't usually have any way to create titles, backgrounds or objects but they give you many more ways to display and manipulate those titles, backgrounds and objects once they are created with other programs.

**Text/Screen/ Display Programs**

These programs perform most of the functions that a character generator does. They let you create a series of text/screens and then give you a number of ways to display them too. Usually they let you store a number of screens that you can then play back. Sometimes you are limited to the fonts that are supplied with the program and sometimes you can bring in fonts and backgrounds from other programs. These programs also give you a wide variety of transition effects to choose from when you want to display these screens. While they are not quite as versatile as the other two kinds of programs they are self contained, easy to use, fast and produce very good titles.

The two things that are important about titling are the text and the transitions. Right now there are dozens of fonts designed specifically for video applications and I'm sure there are more on the way. There are also hundreds of transitions possible with the various titling and presentation packages. Here are some things you should know about each of these.

# Text in Titling

While the Amiga has a number of fonts included with the system, not all of them are appropriate for video work. The first problem with the Amiga fonts is that most of them are too small for video. Turn on your Amiga and step back from the screen a few feet and you will see (or not see) that the text that appears automatically is not large enough to read clearly except when you are seated directly in front of the monitor. In the early days of desktop video some cable companies used Commodore VIC-20s because they displayed 22 characters on a line and that was just about perfect for video. There are no hard and fast rules about how small text can be on a television and still be readable. Unless you are trying to hide information in "the fine print" at the bottom of the screen you will want your text readable by the viewers. So the first step in choosing the right text is to pick a font that is large enough.

A font is just an alphabet (usually with some additional punctuation marks). The point size is how large the text will appear. When you talk about point sizes in print media it is fairly exact, but when you start talking about point sizes for fonts in video things get a little vague. The best way to judge if a point size is too small is to put it on a screen, step back about six to ten feet, and if you can still read it then it should be fine.

Right now there are three types of fonts and not all titling programs can use all three. The first type are the fonts that come with your Amiga. Naturally enough they are called Amiga Fonts. The second type are custom fonts. These are specially created fonts not included with the Amiga. They are usually available for a small price or even for free (public domain software). Any font that isn't an Amiga font is considered a custom font. The third type is called Color fonts. These are special fonts that have their own custom color palettes linked to them when they are saved on disk. Sometimes these color fonts have been designed for color cycling effects. A handful of titling programs can handle color fonts created with other programs. Color fonts can be very impressive however, in order to get their full effect you must use their particular color palette which may effect the rest of your text/screen.

If none of the titling software or font packages give you exactly what you want, there are programs that will let you design your own fonts. There are also outline fonts available where you fill them with whatever color or image you want. Designing your own fonts can be fun but they take time and a certain amount of creative ability.

## Character, Line and Screen Options

Most titling packages offer a number of ways to modify the fonts, lines and backgrounds you use. The main options for individual characters are color, cycling, flashing, italic, bold, underline, outlining, shadowing, extruding and anti-aliasing. The options for entire lines of text are justification, positioning, kerning, background and transition. The options for entire screens are usually limited to background and transition.

### Character Options

Color — Just about every titling package lets you modify the color of individual characters. The only things that you need to know about character colors is to avoid "hot", over saturated colors (stick to off white or grey rather than pure white, etc.) and if you are using color fonts it is best to load those fonts first and stick to that palette for the rest of your screen. You might even want to stick to that palette for all your screens as transitions between screens that use different palettes can cause your colors to shift in the middle of a transition. If you really want to change palettes between screens then you might want to pick a transition where the first screen fades to black before bringing on the second screen.

**Cycling**

There are two forms of font color cycling. You can either cycle the colors of the fonts the way you would any graphic (using the palette and cycle controls in the titling package) or you can cycle the character colors of a color font. The difference is that most titling packages only allow one color for the characters so that the entire character will change colors in cycle mode. With color fonts (where the colors of the characters are pre-defined by their designer) or multi-color fonts (where the colors are defined by the user in the titling program) the colors within each character will cycle.

**Flashing**

Sometimes this effect is called blink but it is just a form of color cycling where character(s) blink on and off or from one color to another. This is useful when you want to draw attention to a particular word on a title screen (for example, the word SALE flashing on and off).

**Italic, Bold, Underline**

These are just what you would expect them to be. Italics slant the characters, bold makes them thicker and underline draws a line under the character(s). Usually, the only one of these features that you can adjust is the underline. Depending on the program you can define how thick an underline you wish and where you want to place it relative to the characters.

**Outlining**

Outlining is simply drawing an outline around each character, usually in a different color than the character color. This is very useful in video work where text might be overlayed on many different colors. If you have ever watched a foreign film with bad subtitles you notice that sometimes the subtitles are fine but on some scenes the white letters are unreadable on a white background. Outlining helps make the characters readable no matter what color the background is. Unless you are creating your text and backgrounds carefully it is a good idea to use the outlining features for all your text.

**Shadowing**

Shadowing is just what it sounds like; putting shadows behind the letters. There are usually two variations on shadowing letters, a filled shadow or a cast shadow. Filled shadows run from the edge of a character off at an angle. This is almost a 3D effect. A cast shadow gives the appearance of the character floating above the background casting a shadow below. Most titling packages give you control on how much shadowing you add to your text.

**Extruding**

Extruding is a 3D effect, as if the characters were carved from a block. Some titling packages only do a pseudo-extrusion where they simply use the character color to create a filled shadow. If the characters aren't viewed from different angles this type of pseudo-extrusion works fine but in some special presentation packages you would notice that the characters are not truly three dimensional.

**Anti-Aliasing**  Finally, there is anti-aliasing. This is a video/computer technique for making characters easier to read on a television screen. The computer looks at the edges of the character and compares them to the background, it then draws a border around the character using a color that is halfway between the character color and the background color. Anti-aliasing smooths rough edges and reduces "jaggies." Many fonts that are included with titling packages are anti-aliased already (as if they were placed on a dark or neutral background).

## Line Options

**Justification**  This is a handy feature for lining up your text on the screen. You usually have four justification options which are; left, center, right or free. Left justified text will start each line neatly along the left side of the screen, right justified will cause each line to end neatly along the right edge of the screen and center justified will make your text look nice and centered in the middle of the screen. Free isn't really a justification mode it is more like a lack of justification. You place the text anywhere you want on the screen. Some word processors will offer flush justification (where the beginning and ending of each line of text lines up neatly, like a newspaper or book) but this is not an advantage when it comes to video titling. Word processors line up the text on both sides by adding extra spaces between words and letters which can look awkward on a television screen. There is one other form of justification offered by a few titling packages and that is grid snap. While you are composing a screen of text a grid of lines is displayed on the screen. Text is treated like a brush that you can move around with the mouse. When you place your text near any of these lines and click the mouse button the program will automatically "snap" your text to the closest grid line. This is handy for creating charts and tables.

**Positioning**  Some titling programs let you put your text anywhere on the screen you want while others will only let you place text within certain lines. In the free style you can overlap text if you want while in the line oriented programs you have less control (but it is easier to get pages of text looking nice and neat). Some line oriented titling programs will, however, let you define a line that is larger than the font size and then adjust the positioning of the text up or down within that line.

**Kerning**  Kerning is the amount of space between letters in a word. If kerning is turned off then every letter takes up the same amount of horizontal space. With kerning turned on the letters are pressed closer together so that a lower case "L" would take less horizontal space than an "R". This means that the word "ill" would take less horizontal space than the word "war". True kerning is sometimes built right into a font so that

| | |
|---|---|
| | each character has a different width and sometimes kerning is estimated by the program. |
| **Line Background** | While you will usually want to create or import a background for the entire screen some of the titling packages let you modify the background on a line-by-line basis. The options for line backgrounds can range from a simple solid colored bar to complicated patterns. Some packages let you put your text in boxes, circles or ovals. Some let you put lines around a word or line of text. The two things to keep in mind are to keep the background subdued or the text will be lost and while a screen may look fine on your computer monitor it might not look that great once it is transferred to video tape. |
| **Line Transitions** | Transitions are a major part of any titling package. There are hundreds of transition effects available for getting from one screen to another. Some programs let you define transitions on a line-by-line basis. This way you can have parts of the screen remain the same while a single line (or lines) changes. In a way presentation programs are nothing but transition programs. They let you decide how things get on the screen, what they do while they are there and finally, how they leave the screen. |
| **Screen Backgrounds** | Almost all of the titling programs let you modify the screen background in a number of ways. You can have the background be color zero when you want to superimpose text on external video. You can have the background be a solid color. You can sometimes bring in IFF pictures for a background which is handy when you want to put titles on top of digitized images. There are also companies that supply backgrounds for video, animation or paint programs created by artists. Many titling programs will let you add or create patterns for your background and some can get very fancy, including things like tiling, wall paper, mirror, even dithering effects. Like the line backgrounds you should try to stick to more subdued background images or your text may get lost. |
| **Screen Transitions** | The two most important features in a titling package are the text and transition options. There are hundreds of transition effects and more are being created every day. All the presentation and text/screen/display programs will give you a number of transition effects and some will let you create or modify your own. While there are many special names for special transition effects here are a few common ones. |
| **Cut or Bang** | The new image instantly appears replacing the old. |
| **Wipe** | The old screen is "erased" revealing the new screen below. Wipes can be in any direction (even from the corners) and in a number of patterns. |

| | |
|---|---|
| **Push** | The new screen "pushes" the old screen away. |
| **Slide On** | The old screen is covered by the new screen sliding over. |
| **Slide Off** | The old screen slides away revealing the new screen. |
| **Flip** | As if the new screen were rotated from a horizontal position (there are dozens of variations of this). |
| **Tumble** | As if both screens were printed on two sides of a sheet and it is turned over to reveal the new screen (there are many variations of this). |
| **Roll or Scroll** | Where the text moves smoothly up (or down) the screen. Almost all movie credits are scrolled by just fast enough so that you can't quite keep up. |
| **Crawl** | Where text moves along the bottom of the screen (usually right to left). |
| **Teletype** | Where each letter pops onto the screen one at a time (as if someone were typing them). |
| **Dissolve** | Where one screen slowly fades away as the new screen appears (at one point both screens are displayed simultaneously). |
| **Fade** | Where a screen fades away (usually to black but it could be to any color or background). |

These are only a few of the more common transition effects. You will have to check your titling software manual for the names of others. You will also want to play with them to find out how they look. One tip on using transitions in a production. While fancy transitions are fun they can also be distracting. Pick one or two transitions and stick to them. You don't have to have a different transition for every screen of text.

## Picking a Text/Screen Program

About the only thing that you have to know about text/screen programs is that they are not stand-alone programs. They are essentially just for creating one screen at a time. You could use these screens one at a time but there are no transition effects built into these programs to let you manipulate how multiple screens are presented.

### TV*TEXT Professional from Zuma Group $169.95

Zuma Group has been a creator of font packages and text programs for a long time. TV*TEXT Professional is a program that lets you create one screen at a time and save it for later use (you could of course use just one screen at a time). It works with Amiga fonts, color fonts and includes a number of fonts with the program. First you select the font and then type in the text string in a window. The text string is then rendered, and then like a brush, you can place the text anywhere on the screen. You can switch between free, justified or grid/snap options. Beyond the normal font choices of color, outlining, shadowing, etc., you can also specify metallic textures, glow, sheen, cycling, strobes, extrusions and a number of other special options. There are a number of paint program features for drawing boxes, circles and ovals around text and you can perform cut, copy and paste operations. You can also take a section of the screen and rotate, flip and re-size it. There are a number of special background creating features like dithering (where there is a gradual color change), tile and wallpaper (where any area can be repeated all over the screen) or you can bring in IFF images for backgrounds. TV*TEXT will let you save the entire screen or parts of the screen as objects. As far as flexibility, this program gives you many more options than most other titling packages. Its main advantage is in creating custom titles.

### Video Effects 3D from Innovision Technology $199.95

This program could be considered a kind of one-shot presentation package but its main function is to create a single text/screen effect. Starting with an object (created with another program) Video Effects gives you unlimited options as to how you want the object manipulated as it appears on screen, what it does while there and how it leaves the screen. This program does true extrusions of text so that as the object is manipulated in 3D space you see all the sides. Once an object is extruded you describe its movements around the screen. You can then preview the actions in a wire-frame mode. Once you are satisfied with the actions of the object the program renders all the images in the sequence (60 per second but this is adjustable) and saves them to disk as an ANIM file that can be played back later. This program is particularly effective for creating special flying logos and special text effects. Since you control the movements and rotations (in three dimensions) you can create unlimited transition effects. Since you can only control one sequence at a time it could be considered a text/screen program but it is more than that.

## Choosing a Presentation Program

Presentation programs do more than straight page flipping programs. They can usually play back ANIM files or create them. They can manipulate individual objects and screens. Most of them have features that let you incorporate digitized music and sounds. They have numbers of transition effects and many let you create your own special transitions. While they can create sequences for video titling they can also be used for business presentations or running demos so there are functions you may never use in your studio.

### Deluxe Productions from Electronic Arts — $199.95

This program lets you put together presentations composed of up to 12 scenes with backgrounds (but you can chain productions). Each scene is made up of up to 5 clips. Each clip can be an object, brush or text. You can have the program move the objects with up to 10 points on a path. There are 40 screen transition effects, 19 object transitions (9 'wipe on's, 9 'wipe off's and a 'stay on screen' transition). Deluxe Productions does support overscan and most resolutions.

This program is fairly easy to learn and use. You first select a background IFF picture you want for a scene (including it's transitions) and then what clips (which are objects, brushes or text) you want to manipulate. You then select the objects 'wipe on' transition and starting position anywhere on or off the screen. Using the mouse you mark the points on the screen where you want the object to move, set the speed and the object's 'wipe off' transition. While there are special fonts included the program cannot use them directly, they must be loaded into another program like Deluxe Paint.

### TV*SHOW from Zuma Group — $99.95

This program lets you create scripts that contain 'events'. There are six event types; screen, object, cycle, loop, speech or key. Each event is made up of an event number, filename, type, 'on' transition, dwell time, 'off' transition, delay time, transition speed and cycling on/off. With fifty built-in transition effects and variations you can create a number of unique titling or presentation sequences with TV*SHOW.

Like the other presentation packages TV*SHOW uses screens and objects created with other programs. Composing a script is fairly simple using both mouse and keyboard. One of the few limitations of this program is that only one event at a time can be executed. This shouldn't be too much of a problem unless you are attempting something out of the ordinary (such as a shatter effect where an object breaks into separate pieces).

## The Director from The Right Answers Group $69.95

The Director is a slightly different variation of presentation program. Rather than using the mouse you create scripts with a special programming language that is similar to the BASIC language. This means that the program is a little harder to master but it also gives you more precise control over the finished presentation. It supports all resolutions and includes a number of built-in transition effects such as fades, dissolves, stencils, flips, as well as page flipping and color cycling.

One of the most powerful features of this program is something called a 'blit.' The Amiga has a unique capability to shuffle individual elements of screen data very, very quickly. What this means to the desktop video producer is that you can specify a block of the screen and have the Amiga move the block anywhere on the screen. With the Director you can 'animate' anything (or part of anything) that appears on the screen, from a simple movement all the way to complicated multi-level effects (like shatters).

## Lights, Camera, Action from Sparta $79.95

Lights, Camera, Action (LCA) is a presentation program that specializes in IFF images, music and ANIM manipulation. The program contains over 40 transition effects, supports just about all formats and resolutions (including PAL) and lets you link music and sound files to a script. Using the mouse and keyboard you create a script where you can control the IFF graphics, and ANIM file transitions, playback speeds and link music files to the action. Since you can alter the playback size and speed of ANIM files you can do more than just play them back and link them. If you use another program to put text transitions into an ANIM file you can then customize the file. The only restriction is that to manipulate ANIMs they must be saved in severe overscan (384x480 and 768x480 NTSC or 384x600 and 768x600 PAL).

LCA has some unique features that make it good for creating presentations or videos. Its ability to manipulate and create ANIM files beyond just playing them back, and its advanced music and sound options (repeat, pitch, volume, start and stop points, etc.) are quite helpful. Included with the program is a special utility called GrabANIM that you can use to turn any sequence of screens into an ANIM file. The example given in the manual is to enter some text into a paint program, turn it into a brush, grab that as a frame, move the text slightly, grab another frame, etc.

## Choosing a Text/Screen/Display Program

These all-in-one titling packages are very popular in professional video situations because they are simple to learn and use, plus they do a very good job. They don't usually give you quite as many ways to create or modify a screen as you have with a text/screen program or as much control over the transitions as you would get with a presentation program. But if you need to do a lot of titling these programs give you just about everything you need.

### Broadcast Titler from InnoVision            $299.95

This program works in most resolutions and modes, will let you use the supplied high definition anti-aliased fonts, any Amiga fonts or fonts created by other companies, even Color Fonts. You can control the color, outlining, shadows, color cycling, have individual characters flash on and off, even do 3D extrusions. You can store up to 1000 pages of text and there are 70 transition effects to pick from (including continuous credit rolls). Each transition effect has 9 speeds and can be applied to the entire screen or to individual lines of text. The program also offers additional software controls for the SuperGen genlocking device.

Broadcast Titler is easy to learn, easy to use, offers a wide selection of transitions and produces impressive, professional results. This is one of the best all-in-one titling packages that should perform well for both the home video producer and the professional.

### Pro Video Gold from Shereff Systems            $300.00

Pro Video Gold is one of the most widely used titling packages. It works in most resolutions, can store up to 2600 pages of text (with additional RAM), comes with 16 font choices (with more available from the company). You can adjust the font outlining, shadows, color cycling, flash characters on and off and select between 1, 2 or 4 color fonts. The program contains 99 transition effects including continuous roll and crawl.

The program is easy to use (all options are activated with the function keys) and the manual is very well written. It is not surprising that Pro Video is one of the more popular titling packages for professional users. Once you learn the system you can compose screens quickly enough for live situations.

## VideoTitler from Sparta                $149.95

VideoTitler is really two programs in one package. The main program, VideoTitler is a text/screen composing program. The second part, VideoSeg, is a scripting presentation program. VideoTitler features some of the more sophisticated character manipulation options available in any titling program. Beyond supporting just about every resolution and mode the program lets you use any font and includes special 'poly fonts' that can be stretched, mirrored and distorted. This makes VideoTitler one of the only packages that can display text on a diagonal. There are over 20 character manipulation options including edge, outline, neon, star, 3D, shadows, etc., and if that isn't enough there are tools for creating your own effects. You can import IFF graphics and brushes then clip, paste and distort the images.

Once you have created your screens you use VideoSeg to create a slide show script (similar to Lights, Camera, Action described above). You can use VideoSeg to display your sequence or it will create an ANIM file for you. This section of the program doesn't have as many transition options as some of the other titling packages but in a scripting environment you have more control over those options.

## Animation:Titler from Hash Enterprises                $79.95

This program is almost a combination of all of the above types. You can use any font (including Color Fonts), import background images or create your own using a wide variety of screen manipulation techniques. Titler supports just about all resolutions and modes, lets you adjust character color, outline, border, shadow (even shadow color), justification and transition effects. You can also position text anywhere on the screen using the mouse. The program can either play back your complete presentation or create an ANIM file.

One of the big advantages to Animation:Titler is that it will work on a 512K Amiga with only limited restrictions. Editing screen sequences is also very easy. While you may not find quite as many transition effects in Titler as you do in some of the other programs, you can do things like center screen rolls (where the top and bottom of the screen remain and only a few center lines change). Even though the program is mouse and menu driven it is a little more complicated to master. Once you have mastered it though you should be able to perform almost any titling function you need.

## Titling Software Tips

There are two approaches to video titling software. First, you can buy an all-in-one package. The advantage to doing this is that the programs described above all work very well, they are all easy to use and you don't have to worry about mixing and matching software programs, resolutions, palettes, etc. These types of programs are more appealing to professionals because they are more like the family car that is used every day. All they want is a program that does the job consistently well, without a lot of confusing extras. The disadvantages to buying an all-in-one titling package are that you don't have quite as much flexibility in how you create and display your screens. You may also run into problems if you want to mix and match the results with other programs. Finally, these programs usually require more computer memory to run.

The second titling option is to buy a text/screen composing program and a presentation program. You would use the first program to create the desired screen and then use the presentation program to link the screens together (unless you just want to do straight cuts between screens). The advantages to this approach are that both types of programs offer you greater freedom. Also, since you are only creating one screen at a time and the presentation programs usually put everything together in ANIM files, you can get by with less computer memory. The problems with this approach are that the programs are a little harder to master and usually take more time in the long run. Another problem is that you have to be careful about mixing formats and resolutions between programs from different manufacturers.

The ideal set up would be to have both types. For your day-to-day applications use the all-in-one titling packages and when you need something special use the others. If you can't buy both solutions you should think about what your main titling requirements are. If you are going to be doing lots of titling (particularly credits) on many projects you might want to go with a text/screen/display type package. In a pinch you can use your paint program to create a special screen. Keep in mind, however, that paint programs can't usually handle text as well as titling programs. On the other hand if you are going to be doing special title screens (perhaps including custom logos), plan to do live presentations, interactive demonstrations or if you need extra fancy titles mixed with other special effects then you will probably want to get both a text/screen program and a presentation package.

There are a few general rules about titling but nothing is absolute so you should experiment and decide what works best in each case. Here are a few guidelines you should consider when you get into titling.

- As I mentioned before, just because a program offers dozens of transition effects doesn't mean you should use all of them in every situation. Pick one or two and stick with them. You don't want the transitions to detract from the words.

- Try to stick to neutral backgrounds. You don't want a lot of action or complicated images drawing the viewers eyes away from the titles.

- Most titling packages require at least 1 megabyte of RAM or more (Animation:Titler is one exception). Check the package before you buy to be sure you have enough RAM to use all the effects. You will probably want to have 1 to 3 megabytes of RAM in your Amiga eventually.

- Remember the illegal colors problem (discussed in Chapter 6)? Stick to pastel colors and off-white colors rather than fully saturated colors. Bright colors will tend to 'bleed' when transferred to video making the words difficult to read.

- Outlined characters and characters with dark or grey shadows are easier to read on a television screen. This is particularly important if you have varied color background images or if you are genlocking over external video.

- Anti-aliased fonts look better on television screens. Try to use them when you can.

- Some genlocking devices change the clock speed of the Amiga which may effect any of the titling programs. You will have to try things with your particular setup to see if this will be a factor.

- Use the largest point size font you can to make it easier to read each screen.

- Don't try to put too much on the screen at one time. Even though the Amiga screen can display a great deal of text at once remember that you are doing videos not TV books. Viewers will get bored quickly if they have to read a lot on one screen.

- Leave each screen up long enough for a slow reader to get through most of it. There are no hard and fast rules here. A quick rule of thumb is leave a screen up long enough for you to read it twice (since you know the words already you will be able to read it faster than a viewer seeing it for the first time).

- Proofread your screens twice and double check your spelling. This may sound silly but the last thing you want is a stupid mistake in a great looking title screen (particularly if someone else is paying you).

- When everything is set up do a trial tape first and check the results. What may have looked great on the computer screen may look poor on video tape.

## Summary

A titling package is probably the most useful software that a desktop video (or even professional) producer can own. Even simple titles with simple transitions can greatly enhance a video production. With the Amiga and some easy to use programs you can do titling like the professional broadcasters. Titles and credits elevate a video from 'home-movies' to 'home-productions' with just a few words.

# Chapter 9

## Music and Videos

# Chapter 9
# Music and Videos

Music, sound effects and visual entertainment have gone together since the beginning of time. If it was someone imitating a lion while they acted out a story, singing or playing a drum during a puppet show or an orchestra playing behind a stage performance, there was usually something. Even the early silent films would be shown in theaters with a piano player trying to enhance the mood. These days we hear more music than at any time in history. Almost every commercial on television has music and sound effects. Every program has a theme song, every film we go to has background music and sound effects galore. Talk radio stations will play music, shopping malls and stores play music, even elevators have Musac. It would be hard to go through an entire day without hearing a dozen different songs.

If you want to see the impact that music can have in a video production, try video taping a minute or two around your home (assuming that you have a camcorder) with no music, now go back and tape the same scene with the stereo or radio playing different music in the background. Another way to see this is to turn the sound down during a dramatic scene in a horror movie. The scene will suddenly be less frightening.

There are a number of ways to get music and sound effects into a video production depending on what you want, what you have, what you can do and what you are planning. I'll talk about some general elements of audio in a video production and then focus on how the computer comes into each of these methods. In most cases the terms music, sound effects and voice overs are interchangeable. You could consider a sound effect to be a very short, slightly strange musical piece that adds to the visual, voice overs could be though of as a song without notes, and music could be considered a string of melodic sounds. In all of these cases what we are talking about is somehow adding additional sound to a video production. You can always just use the sounds that are recorded during the taping but sooner or later you should think about adding extra sounds, music and voice overs.

# When to Add Extra Sounds

There are really only three times that you can add sound to a video production; during taping, during editing or during playback. Each of these have their own advantages and disadvantages. Each of them have their own unique problems. And each of them are necessary in different situations.

# Adding Sounds During Taping

Unfortunately, the audio is one of the last things that we think about when setting up for a shot. We are usually more concerned with the visual. While it is true that no amount of audio magic will make up for a bad visual it is also nearly impossible to correct poor audio. Ignoring the audio during taping will make your job a lot harder later on. People don't notice good sound recording in a production but they will notice if the sound quality is poor.

The two most important things you should be concerned about during taping, whether you are adding extra audio or not, is to get the cleanest audio you can in the first place and to be aware of background audio that you may be taping inadvertently. Getting the cleanest audio you can in the first place is important because it is almost impossible to clean up poor audio later on without a lot of very expensive equipment. Trying to lip-sync voices back in the studio can be a nightmare. Being aware of background audio is important for the same reason. Once it is on the tape it is nearly impossible to take out sounds you don't want without taking out all the audio. Getting everything as quiet as possible can help in both these cases. A person will sound much better talking in a quiet room than trying to shout over crowds, cars or machinery. Turn off all TVs, radios and other sources of music or noise. You won't be able to remove these sounds later and TV music can add unwanted or inappropriate moods. Just before taping take a moment to listen for unwanted sounds. The one unwanted sound you won't be able to hear by doing this is the wind. Taping outside on a windy day can cause problems even for professional audio engineers. Try putting the microphone inside a small cage (a whisk, box, two food strainers) with thin cloth (sock, cheese cloth, stocking) stretched around it.

There are many times when you might want to add extra sounds during the taping of a production. If your master VCR has audio dub features (the ability to add new audio without effecting the video) then adding extra audio during taping is not quite as important. If your VCRs don't have audio dub features then whenever possible you should think about including extra audio during the initial taping of scenes. The main advantage to adding audio during the taping process is that you can avoid an extra generation down the road.

In a studio situation you can usually plan far enough ahead to prepare extra audio to include during taping. Music, sound effects and voice overs can be recorded ahead of time and mixed in while the camera(s) are running. With a little ingenuity you might even be able to add pre-recorded computer generated audio during a location shoot.

Of course there are a number of problems with trying to add extra audio during the taping process. The logistics of coordinating extra audio equipment with your video equipment is always going to be a problem. Timing of audio and visual events can be frustrating. Audio mixing in the field can be difficult or impossible. If you are working on your own it might be best to look at adding audio during the editing stage rather than trying to juggle many things during a shoot.

## Adding Sounds During Editing

To be more accurate, you will probably not be adding sounds during the actual editing process but just before (on individual scenes) or just after you have completed editing the video. Planning is the main tool that you will use when adding audio during this stage. This is where the computer is going to play the largest role. In the creation of music and sound effects the Amiga can help you compose and play back music. With a sound sampler or audio digitizer (a device that lets you bring sounds into the computer) you can create realistic (or other-worldly) sound effects.

The advantages to adding audio during the editing stage is that you have much more control over the final results. You also can take the time to compose your audio, and you can custom fit your audio to the video. Getting the timing right is much easier when you know exactly how long the scene or scenes are. You can also get special sound effects to happen exactly when you want because you can watch and time scenes before adding the audio.

The biggest problem with adding audio at this stage is that it can mean adding another generation. A VCR that has audio dubbing capabilities will avoid this problem. Even if your VCR does not have audio dub capabilities you may be able to avoid adding a generation on some longer scenes, animations or during credits.

## Adding Sounds During Playback

Trying to add audio during playback is something that you will rarely do but there are times when it is appropriate. During a live performance, demonstration, lecture or special presentation you might have to do your own music or 'voice overs' live. In these cases it is not so much a question of synchronizing the sound to the video it is more a matter of adjusting the video to the sound. Presentation software usually has features that will let you customize productions to make them more interactive and some of them will let you link audio directly to scenes.

## Audio Equipment

The range of audio equipment available is remarkable. You could spend as much money for audio equipment as for video equipment, but you shouldn't have to. With some basic audio equipment and things you probably already have you should be able to add impressive sounds, music and voice overs to your videos. Here are some of the things that you will need or may want to get eventually.

**Microphones**  There are dozens of kinds of microphones for different situations and you can break them down into a number of categories. For example, you could divide microphones into the ways they operate. Condenser, Electret condenser, Dynamic, Crystal and Ceramic, Carbon and Pressure Zone Microphones use different internal mechanisms to pick up sound. Each have advantages and disadvantages. **Condenser** mikes have the best sensitivity but are expensive, fragile and require a power source or battery. **Electret condenser** mikes are also good (particularly for voices) but also require a battery and are sensitive to heat and humidity. **Dynamic** mikes are durable and don't need a power source but they are less sensitive to sounds. **Crystal and Ceramic** mikes are very inexpensive but they aren't very sensitive and they are fragile. **Carbon** mikes are those that used to be in telephones, they are good only for speech (and not very good at that either) but they are nearly indestructible. **Pressure Zone Microphones** (PZMs) use the surface that they rest upon as a sound board. Connected to a table they will pick up all the people talking around the table but also finger tapping, paper rustling and knees banging.

Another way to categorize microphones would be according to the directions they pick up sounds. Shotgun or Zoom, Directional or Unidirectional, Bidirectional, Cardioid and Omnidirectional. **Shotgun or Zoom** mikes are very directional, only picking up sounds from the direction that they are pointed. **Directional and Unidirectional** mikes mainly pick up sounds in the direction they are pointed but there will be some 'leaking.' **Bidirectional** mikes pick up sounds in two directions (front and back). **Cardioid** mikes pick up sounds from the front and sides but not from behind. **Omnidirectional** mikes pick up sounds from all directions (these are the kind usually built into camcorders).

You could also divide the microphones into categories depending on how they are mounted. Hand-held, Boom, Lavalier, Parabolic, Radio, Telephoto and Stand microphones are all examples of different mountings. **Hand-held** is obviously a mike that is held by someone, you have to be careful about pointing it in the right direction without rustling cables, knocking into things or fidgeting. **Boom** mikes are put on the end of a long pole usually suspending the mike over the subject or subjects. Again, the operator has to avoid knocking into things while pointing at the speaker, and keeping the mike out of camera sight. **Lavalier or Clip on** mikes are either on a string around a speakers neck or clipped to their clothing. The speaker mainly has to avoid moving too much as clothing rustles are picked up. **Parabolic** mikes are in the center of a dish that focus the sounds coming in (like cupping your hand behind your ear). The problem is keeping the dish pointed in the right direction plus the sounds can be distorted somewhat. **Radio** microphones are connected to a small transmitter with a matching receiver at the camcorder or VCR. These mikes are very distance sensitive, can be expensive and are subject to radio interference. **Telephoto** mikes are essentially adjustable shotgun mikes. Some of them can even be connected to the lens of a camera so that they zoom when the lens zooms. They are moderately expensive and can pick up camcorder vibrations. **Stand** mikes are any microphone mounted on a pedestal. They are good for avoiding rustles, banging into things or fidgeting fingers but they are not very mobile so the speaker must adjust their distance.

Another thing you will hear when you start talking about microphones is **high impedance** and **low impedance**. While it won't really hurt things to plug a high impedance mike into a low impedance mixer (or vice versa) it won't sound very good either. Just try to match your audio equipment to impedance levels.

**Which Mike to Buy**

As far as what type and how many microphones you will need depends on your application. It will only take a short time before you realize that the microphone built into your camcorder is not going to do all the things that you need. For voice overs you will need a stand mike. A Unidirectional, Cardioid or Omnidirectional Dynamic mike with a table stand should work fine for this. This mike can also double as a hand-held mike for interviews and studio work. The camera mounted telephoto mike can improve the quality of your camcorder audio in a number of situations. These are probably the first mikes that you should consider buying. If you are going to be doing lots of interviews or studio talk show type programs then you might want to get a few Electret condenser lavalier microphones (most of them come with clip on attachments). Unless you have a crew to help out you probably won't get into boom mikes. If you can afford them radio mikes are good for outdoor shooting, particularly for long shots.

**Mixers**

The next piece of audio equipment that you will need is a mixer. A mixer lets you blend multiple audio sources into one. The reason that you need a mixer is that if you want to put voice overs, music or sound effects on top of the audio from the original tape you have to mix both the audio from the tape and audio from another source. Now, since the audio reproduction in video is not even close to the audio that comes from a CD player you don't have to buy top-of-the-line equipment (which can run into thousands of dollars), but you will need a mixer with enough inputs to accommodate all your sources. At a minimum you will need a mixer that will accept three inputs; an input from your slave VCR (the original audio), the voice over microphone and the music. Eventually, all VCRs will be in stereo but you don't really need two channels for each of these sources. You can get by with fewer inputs if you have to but you will end up doing a lot of cable switching, recording and re-recording. Radio Shack has a few inexpensive mixers that will work just fine for a desktop video situation.

Audio mixing is almost as complicated as video editing and there are people who get paid a great deal of money for their audio mixing skills. You will probably want to do your audio mixing off-line. In other words, you will want to put your music and even voice overs on audio tape first, before you transfer them to video. Which brings up the next piece of audio equipment you will probably want.

**Tape Decks**

While you could try and coordinate all your audio sources at the same time it is much, much easier to pre-record as many pieces as you can ahead of time. If you have ever tried to read a few lines of a script out loud you know that it is very difficult to get it right. A tape deck is almost essential for recording music and voice overs. It doesn't have to be a fancy deck, all it has to do is record and play back with a reasonable quality. The odds are that you already have a tape deck but if you don't there are a number of very inexpensive decks that can do the job. If you can afford a second tape deck (even a very, very cheap one just for voice) it can make your life a lot easier. Even though there is degradation of the audio as you dub from tape to tape the quality of TV audio is so much lower no one should notice (unless you have added a great deal of 'hiss' or 'hum').

### Amplifiers and Speakers

You probably already have a stereo with an amplifier, turntable and speakers (probably a CD player and tuner as well). Obviously, you are going to want to hear the results of all your audio mixing and recording somehow before you put it all on video tape and this requires an amplifier and speaker(s). In most professional video studios you will find a very cheap TV set and a very cheap speaker (along with the expensive stuff). The reason that they have these is that they want to see and hear what the production will be like under the worse possible circumstances. You don't have to have a super-high quality audio system for video applications but you want to try to maintain the best audio you can. If you want to check how your audio will sound before you record to video you can feed it into the master VCR, put the VCR in record/pause (with a scrap tape) and listen through your TV or monitor speaker. This will give you a good idea how it will sound but you should do some real recording to video tape to hear the final results.

### Cables

There is no way around it, you will need a lot of cables in your studio. As far as audio and video cables go, you can use video cables for audio but you probably don't want to use audio cables for video. No matter what kind of cables you get (or make) you should get shielded cables. It might not make much difference to your audio to use unshielded cables but it can make a difference to the video equipment that is next to it.

## Computer Audio

There are two primary uses for computers in audio; music and sound effects. While the Amiga does have a built-in speech synthesizer it is far from being able to realistically simulate a human voice. If you need a voice that is obviously a computer talking (a re-make of 2001?) then the Amiga speech is very simple to use. Just read the section of your Amiga manual covering the Say Speech Synthesizer. The one exception to this is when you digitize voices, and I will talk about them later. Apart from special circumstances you will mainly be using the Amiga for music and sound effects.

If you are a musician then you may already be interested in computer music. If you are not really interested in creating computer music then you might be thinking about skipping this section but there is a very good reason why you should read on. In Chapter 4 I talked about problems you may encounter with copyrights on images taken off the air. The same problem pops up with music. Every record album, every tape, every CD, every song you hear on the radio is copyrighted by someone. If you want to use music created by someone else then you should get permission or only show your productions to yourself and your closest friends. You can buy sets of albums with non-copyrighted music but they are expensive, out of date and for the most part pretty dreadful stuff. On the other hand there are records and tapes with thousands of sound effects that are not expensive and can save you a lot of time.

Even if you have little or no musical abilities the computer can help you create reasonably good music and you can customize it to fit your needs. If, however, you do have musical skills then the computer can be your personal orchestra. The Amiga is one of the better computers for producing sounds and music. It has special chips that process audio information and it outputs stereo sound. With the right software and hardware you should be able to produce professional quality music and have some fun doing it.

The kinds of hardware and software you will want depends on how much you want to spend (both time and money) and how good a composer and musician you are. Computer music in a video situation is one of the few areas where more money does not necessarily mean better quality. In fact some of the least expensive music software is the easiest to use for the novice. The more complicated you want to get the more money you will have to spend. But even if you end up spending a small fortune for music software and hardware you are still limited by the low audio quality of the video tape and television speakers.

## Music Software

In order to get the Amiga to make music you need software. The music software available is quite extensive and ranges from the very simple to the very complex. Most of the software lets you compose and play back through the computer but some of the more advanced music packages are mainly for manipulating **MIDI** devices (Musical Instrument Digital Interface) or using sound samplers (a hardware device for getting sounds into the computer, much like the image digitizers discussed in Chapter 4). Unless you are serious about music composition and live performances you will probably be more interested in the less sophisticated music software.

There are a few music programs for the Amiga that are not considered "serious" music programs but they can be ideal for desktop video. They let you put together songs quickly and easily, even if you are tone deaf. In a way they are song generating programs. You select a type of music, the tempo and the instruments you want to use. The program then lets you use the mouse or keyboard to 'play' along with pre-set rhythms. The nice thing about these programs is that they are idiot-proof. In other words, you couldn't play a sour or off-key note if you wanted to. As soon as you select one of the pre-sets all the wrong notes are disabled. While you may not be able to compose award winning music with one of these programs, they are great for the musically illiterate. If this is the kind of program that you think you want then there are only a few choices.

### Instant Music from Electronic Arts $49.95

This program does exactly what its title implies. With no skill whatsoever you can create songs within minutes. Even if you can't hum a tune or tap your feet you should be able to create music. The program has a number of built-in song styles from pop to classical and there are a number of built-in instruments you can select from. The templates that you pick automatically disable the 'wrong' notes and keep you on the beat. You decide which of the four voices you want to 'play' and moving the mouse up and down controls the pitch. While the results may not be Grammy Award winning material, it is professional enough for background music.

Of all the mistake-proof music programs, Instant Music is the easiest to use and produces the best results.

The next step up in music software are the note editing packages. These give you much more control over the composition of music and playback. They also let you make your own mistakes. In a way you could think of these programs as word processors for music. Using the mouse and keyboard you enter individual notes on a musical staff. When you are finished you can play the song back. These programs also give you a number of tools for editing music. Some of them will also let you edit sounds. This is a great way to create special or weird sound effects but if you are trying to create a specific sound it could take a while to come up with it (like years).

If you know your way around a musical staff then you might want to go for one of these programs. If you know what you want the tune to be and you know about notation, sharps, flats, rests, etc., then this is the way you will want to go.

## DNA Music, Fractal Music and Protein Music from Silver Software $19.95 each

These three programs will also generate music automatically for you without a lot of musical abilities. They take mathematical descriptions of DNA, Fractals and Proteins and combined with music theory produce songs. The idea may be a little strange but it works.

## Deluxe Music Construction Set from Electronic Arts $99.95

This is probably the best of the note editing programs. You compose your songs using standard music notation on eight separate staves. Entering notes is done with the mouse and keyboard. You can control the loudness, timing and instrumentation. You can modify the sounds of instruments and you can control MIDI devices. While the program is fairly easy to use you still have to know what you are doing in order to create good pieces and even then it will take a fair amount of time and practice.

## Sonix from Aegis $79.95

Sonix offers a more powerful instrument editor than Deluxe Music but it is a little more awkward when it comes to entering music. You still enter individual notes on a staff and the program gives you a number of tools for editing these notes (cut, copy, paste, delete, etc.). Sonix also has one of the best manuals that will not only guide you through the program but give you a quick course in musical notation. You can either use the built-in instruments or using the powerful synthesizer tools create your own. Sonix lets you use the Amiga to play back your creations or control MIDI devices.

**MIDI software**  Beyond note editors you start getting into the hard-core MIDI software. These more advanced programs are not for the casual user. They are for the serious musician. Along with the software you have to buy a MIDI interface device, which lets you connect MIDI instruments like electronic keyboards and synthesizers to the computer. You also have to have a MIDI instrument to plug into the MIDI interface. With the right software, MIDI interface and MIDI instrument connected you can enter music directly into the computer with the MIDI instrument or with the mouse and keyboard. You can have the computer play back the music, control the instruments so they play back the music or both. There are even programs that use AI (artificial intelligence) to adjust the play back according to what a musician is doing live. These programs, interfaces and instruments are far beyond the scope of this book but if you are interested there are many magazines and books on the subject.

## Sound Samplers

One side track in the advanced music area that you might be interested in for video is sound sampling. A sound sampler (and software) is a device that takes an audio source and digitizes it so that the computer can manipulate it. With a sound sampler you can take any noise and turn it into an instrument. A flugelhorn, a human voice, a dog bark, a breaking glass or any sound can be sampled and then modified, distorted or turned into an instrument for use in one of the music programs. This is also the way that some companies get very realistic voices and sounds into their programs. Sound samplers and their software are a little difficult to master but if you are not trying to do highly complicated sound manipulation then you should be able use one in a short time.

### PerfectSound Digitizer from SunRize Industries         $90.00

Everything that I have heard points to this digitizer as being the best of the bunch. When used with sound processing software (like AudioMaster from Aegis or StudioMagic from SunRize) this piece of hardware can produce very good sampled sounds.

The thing to remember about samplers is that the more accurate you want the sound the more memory you need. A few seconds of sound can take megabytes of memory. Human speech doesn't require as much memory as music but you won't be able to digitize more than a minute or two at a time. The biggest advantage to digitizing sounds is that with the right software (Deluxe Video, MovieSetter, the Director, etc.) you can have the computer link your audio and video precisely. Once the audio is digitized the computer can treat it like any other data. Bringing in a digitized sound or sequence is the same as bringing in a new graphic or ANIM file. If you want to add sounds and voices to an animation, a digitizer can eliminate hours of trial and error synchronizing plus you can avoid extra generations.

**Animation:Soundtrack from Hash Enterprises              $119.95**

This is one of the stand out programs for combining digitized audio and animations. It is a powerful package that can be used with a digitizer to sample sounds. Once the sounds are sampled you can manipulate them with an editor. When the sounds are the way you want them you can create scripts that tie the sounds to the visuals. Timing, looping, volume, mixing and other features make the program a powerful tool. The manual is also filled with information about combining audio with animations.

## Audio Tips

There are a thousand things to learn about audio and many of them can be learned from books. But for the most part you will end up picking up knowledge on your own by trial and error. Here are a few of the more obvious tips.

- Pre-record as much as you can. It is much easier to work with recorded audio when you start mixing it with video. Video editing is tricky enough, you don't have to complicate things more by worrying about assembling the audio at the same time.

- Understand the audio cliches and use them, don't fight them. To the viewer tense music means that something bad is going to happen. Romantic music probably won't work during a chase scene. Match the music with the mood. Don't have 'spooky' music during a love scene unless you have a good reason. It confuses the viewer when the music doesn't match.

- Use the music to set the mood. You can, however, turn a scene into something else with the right music. You can change that same love scene into a date with Jack the Ripper with the same 'spooky' music. With the right music you can turn almost any scene into something else. Just be sure that it makes sense to the viewer sooner or later.

- Never let the music or sound effects drown out the speakers. Even when you do it on purpose it is annoying for the viewers.

- If possible, try not to have the audio and video cut at the same instant. For some reason, if you edit the sound first and then the picture a second or two later (or vice versa), people won't notice the edit as much.

- Try to keep the same audio level throughout your production. You can have changes in the volume but not too much.

- Time everything down to the second and then put together the audio. It is usually easier to get the audio to a certain length than to go out and shoot more video. One of the handy things about music programs is that you can alter the speed of playback if you need to fit a piece in.

- Keep it simple. You don't need music and sound effects in every scene and a simple tune can be very effective.

- Watch and listen to movies and television. You will learn a lot by paying attention to what the professionals have done.

# Summary

There is as much to audio as any other aspect of video production and there isn't any way to cover it all in one chapter. While the computer may not play a major role in desktop video it can save you some time, money or law suits when it comes to background music. Music software can help you create songs and play them back. But apart from the few programs that are 'idiot-proof' no software can make you a great composer.

# Chapter 10

## Special Effects

# Chapter 10
# Special Effects

There are an infinite number of special effects (F/X in movie lingo) that you can create with the computer-video combination. Some of them will be subtle and some of them quite extravagant. Most of the F/X you will discover by yourself while trying to get the equipment to work the right way. It is a good idea to jot down some of the things you will run across as you try to digitize, genlock images, animate objects, put in titles, create music, etc. Inevitably you will see things that don't look right for most purposes but could be used for F/X somewhere else. Some of them will look pretty strange and some can be beautiful.

In most cases you will be striving for a particular effect and might run into a dozen others along the way. You could consider anything that isn't straight video to be a special effect. Titles, fades, freeze frames, even cuts might be thought of as special F/X. There are books and books about special effects. There are people who make a living doing nothing but special effects. There are Academy Awards for special effects. Lucas and Spielberg spend millions of dollars on special effects. The main ingredient for creating special effects is your imagination. Let's take a look at the kinds of things that can be done with each piece of equipment in the desktop video studio. I can't go into all the possible special effects you can create so I'll just mention a hundred or so.

## Traditional F/X

If you are really serious about creating spectacular special effects then you should look into the traditional methods used on the stage and for film. Each area has its tricks, techniques, tools and experts. You could spend a lifetime just learning make-up effects. Special film effects, stunts and camera magic are fascinating areas worth exploring but the thing you should keep in mind is to not get too carried away with F/X. Used in the right spots they can be effective (pun intended), but overused and they will detract from the story. Any bookstore or library should have books of information on traditional special effects (check under cinematography, trick) and you can explore them as far as your heart (and wallet) will take you.

## Camcorder F/X

Depending on the camcorder that you own you can do a number of special effects before you get back to the editing studio. Many camcorders have F/X built right into them and some of them have features and capabilities you would not be able to reproduce in any other way. Obviously you should read through your camcorder's manual to find out what it can do and how it is supposed to be done. Here are a few ideas for effects that might not be covered in the manuals.

**Focus**  Adjusting the focus manually can give you a number of special effects. Start a scene focused on a flower in the foreground and then focus on people having a picnic and you have set the mood. If you can match the light levels you can simulate a dissolve by un-focusing on one shot, cut to another un-focused shot and gradually bring it back in focus. A scene that goes in and out of focus rapidly could show a person is intoxicated.

**Camera Angles**  You can generate a number of special effects with camera angles alone. In the Batman television series every time there was a villain in a shot the camera was tilted at an angle. Instead of showing someone hiding under a bed put the camera under the bed and only show people's feet walking by. A camera approaching a house at night and looking in the window can create the effect of an animal, monster or person lurking. Imagine the camera is the eyes of a character, animal, creature or even an object (like a woman putting on lipstick while looking directly at the camera pretending that it is a mirror). Rocking the camera can give the effect of being at sea. Shaking the camera can be an earthquake. Dropping the camera can be a person falling down (or a video producer crying over $1,500 worth of broken camcorder, if you aren't careful).

**Cuts**  The most important transition effect at your disposal is the cut. This is the way that you piece together the parts of a story. You can also use cuts to create effects. Lots of quick cuts can signify confusion. A fist approaching the camera and cut to black might be a knockout punch. Cut from a person driving to a shot of the tire, back to the driver, back to the tire, back to the driver and your viewers will know that there is about to be a blowout. This is also called foreshadowing.

One of the very first special effects used nothing but cuts. With the camera on a tripod have your actors freeze (as best they can), pause the camera and then have one of them leave the scene. Now un-pause the camera and when it is played back, the person just disappears. This effect is no longer 'magic' to the viewer but consider that I Dream of Jeannie, Bewitched and dozens of other television shows ran for years using nothing more than this technique.

**Zooms**

Just about every camcorder available these days has a zoom lens. They are great fun and can be essential for framing a shot without moving the whole camera. You can also create special effects with the zoom. When you first get your camcorder go out and zoom like crazy to get it out of your system because the number one mistake most home videographers make is using the zoom too much. Zooming occasionally is fine, zooming all the time will make your viewers nauseous. Only zoom when it is called for. Once you understand that you can use the zoom creatively your productions will be more effective. Zoom out from people to create shock or alienation. A quick zoom in on people throwing up their arms and screaming in a car followed by a quick zoom in on a tree by the side of the road (combined with the appropriate crashing sound) and you have instant car accident.

**Fades**

Most camcorders these days will let you fade in or out or both. Sometimes to black sometimes to white. A fade can give you nice smooth transitions between scenes. You should keep in mind that a fade has a certain psychological meaning for viewers. A fade is usually used for two purposes. First, to indicate the conclusion of a scene even if the scene is made up of multiple shots. When the actors have completed all their lines or all the information for that scene has been given a fade indicates to the viewer that an entirely new scene is coming up. Second, a fade can indicate the passage of time. For example, if you were doing a travel video you might have a scene of everyone getting into the car and driving away, fade out then back in on a shot of the kids in the back seat, fade out then back in to a shot of the scenery going past the window, then fade out and back in to the car arriving at the destination. The fades between scenes imply that time has passed. (Dissolves also have the same implication).

If your equipment does not have flying erase heads, letting you do clean cuts, you might think about using fades between scenes because no matter what the implication might be for the viewer a smooth fade is better than a glitch. If your camcorder doesn't have flying erase heads but your master VCR does then you would be better off leaving a few seconds at the end of a shot and a few seconds at the beginning of the next shot. Later, in the editing, you can do straight cuts. Don't over-use fades. Try to use them only in the appropriate situations. It would be slightly confusing if you fade out from one shot in a scene and fade in to the next shot in the same scene. If you watch television or films carefully, you will see that straight cuts are used about 90% of the time.

**White Levels**  By adjusting the white level slightly high a beach scene on a cloudy day can still look hot and sunny. At the other extreme by lowering the white level you can make a cloudy day look like night or a sunny day look overcast.

**Time Lapse F/X**  If your camera can do it there are three uses for the time lapse feature. First, when you want to speed up any very slow action like flowers opening, buildings being erected in a few seconds, clouds zipping by, night falling, etc.

Second, when you want to speed up normal actions, like making a car appear to be going 80mph when it is really only going 20mph. Most camcorders won't give you fine enough controls to just speed things up a little but if yours does then remember, there are certain visual clues people will notice when something is artificially accelerated so this trick can backfire on you. Rather than creating a sense of dangerous speed it can just look humorous.

Third, you can use the time lapse feature for animation. Claymation and silhouette animation are done this way. Claymation uses poseable figures (originally made out of clay on a wire frame). The most famous of these were King Kong and Gumby. Silhouette animation uses flat cut-out figures like paper dolls with movable arms and legs. The most famous of these were Clutch Cargo and Johnny Quest. In both forms you shoot a few frames of an object in one position, move the object slightly, shoot a few more frames, move the object, etc. The effect when played back at normal speed is that the object is moving by itself.

You can also shoot animate objects like people (driving down the road without a car for example) by taking a shot, moving a few inches, taking another shot, etc. The effect can be a little jerky, particularly if you are trying this outdoors (with trees blowing in the wind). If you are interested in this type of animation then there are whole books written about it. Usually with a little trial and error you can get a feel for how far to move things.

**Freeze Frame F/X**  There are a few camcorders that will let you freeze a frame in the camera while letting the audio continue to be recorded. This can be a great help in simulating dissolves (see Chapter 5).

**High Speed Shutters**

About the only special effect you can create with high speed shutters is cleanly freezing fast action later in the editing studio. If you are doing a lot of sports taping then a high speed shutter is a must. You should be aware that to use the high speed shutter settings you need plenty of light. Also, if you use the high speed shutter settings for normal action there can be a little jerkiness.

**Strobe**

Some camcorders offer a strobe effect feature that records a series of freeze frames. The effect is a little limited but you could use it effectively for simulating a disco dance floor, a montage of a fight or someone on drugs.

**Voice Overs**

Normally, you would do your voice overs back in the editing studio where you have time and quiet but sometimes you can do your voice overs while shooting. It is always a good idea to rehearse the lines before you begin taping. Voice overs during taping are also a good way to mark your scenes for later reference. A quick and easy way to do this is simply by stating the time and date or scene number after starting the camcorder but before the action begins. It can also be useful to describe what you are taping if you know that later you will be doing a clean version of the voice over.

**Close Ups**

Macro lenses can be useful for creating an opening sequence or titles. By taping still photographs or hand writing the titles (with a nice dark marker on a grey piece of paper), you can put together a nice introduction to any video. You can also use extreme close-ups of household objects to set a scene. Imagine a peaceful family room shot then a very tight shot of a knife blade, back to the family room, back to the knife being picked up, back to the family room, tight shot of the empty counter top. You get the idea. You can also try the old monster movie technique of tight close-ups on insects or animals running, hopping or slithering through a miniature city (lizards are the traditional stand-ins for dinosaurs). And just about every outerspace film or TV show uses very detailed spaceship models shot in extreme closeup.

**Filters**

There are dozens of filters available for still cameras that you can use in video to create effects. There are star filters, colored filters, graduated filters, prism filters, center focus filters and many others. The best thing is to go down to a camera shop and ask about them. The only thing you have to be sure of is that your camcorder will accept that size filter.

If you want to experiment you can try creating your own filters (be careful about damaging your lens though). Cheese cloth, nylon stockings, screens, colored plastic, kaleidoscopes, mirrors, petroleum jelly, cardboard cutouts, almost anything that is transparent, translucent or does something to light can be tried in front of the camcorder. You just have to be careful that you don't block the infrared auto focus sensor (of course you could always focus manually). You can also use reflecting surfaces as a form of filter. One way to get a ghost effect is by shooting through an angled sheet of glass. With the right lighting you will only get a partial, ghostly image from the reflection and a clear image straight through the glass.

**Mis-adjustment** For the really strange effects there is always mis-adjusting the camcorder on purpose. Most camcorders are designed to keep you from accidentally making mistakes like this but you can sometimes put everything on manual and then play around.

# VCR F/X

Like the camcorder, most VCRs are designed to prevent you from making mistakes during recording but there are a few things that you can do to create special effects. Depending on the features built into your VCR(s) you may or may not be able to do too much but here are a few ideas.

**Editing** As I said before, you could think of anything that isn't a straight shot to be a form of special effect. If you look at it that way then 99% of your VCR F/X will be on how you edit your tapes. The final sequence of your shots is the special effect. Taking miles of tape and turning it into a story is where the skill and talent comes in. When you watch a film or television show carefully you may find that many of the special effects are done just by clever editing. In the famous shower scene from Alfred Hitchcock's Psycho at no time do you see anyone actually being stabbed. There are screams, the knife flashing by, blood in the drain, etc., but no stabbing.

The right cuts can create a variety of moods, atmosphere, foreshadowing, flash-backs, and on and on. There are dozens of things to learn about using straight cuts in a production and most of them only come with experience. A few of the more obvious rules are don't cut from one type of shot to the same type of shot. In other words, don't cut from a 3/4 left profile shot to another 3/4 left profile shot of the same person. If two people are talking with each other, one should be a left profile and the other a right profile. If something moves off screen to the left or right don't cut to the same object or person traveling in the opposite direction. There are lots of other things like this that you won't discover until you edit some shots together and then play the scene back.

**Glitches**   Although you should try and avoid glitches at all costs there are rare times when you might actually be able to use glitches in a special effect. In the movie Alien they made use of VCR video glitches to simulate otherworldly interference. As the astronauts enter the alien ship their helmet mounted cameras begin failing (indicated by the glitches) and eventually they die (both the cameras and the astronauts).

**Speed Changes**   If your slave VCR can playback in slow motion or fast motion you may be able to record scenes at speeds other than they were recorded at. You may find, however, that the picture is too unstable to record properly at the wrong speed. Double speed scenes of people packing a car, doing housework, eating, gardening, playing in the yard, etc., can provide a fairly humorous special effect (and can get your point across without boring people with an overly long sequence). The same comic effect can be accomplished by playing scenes backward.

**Over Dubbing**   Over dubbing falls into the same category as glitches used as special effects. Most of the time you want to avoid extra generations. But if you want to simulate a space warp, force field, futuristic weapon blast or characters viewed through the eyes of an alien being or drug addict try re-dubbing the scene five or six times (you may have to go further if you have better equipment). The picture quality will deteriorate, the colors will bleed, and if you go far enough everything will certainly look pretty strange.

# Genlocking F/X

Genlocking is a special effect to begin with so you don't have to go too far to create something special or unusual. Genlocking is also the main way to get the computer effects to be more than just images on a monitor and turn them into part of a video production.

**Overlays**   Overlaying anything from the computer onto video is a special effect according to professional video people. Most of the time you will be trying to do it straight, in other words, with as little distortion as possible. You can also combine a straight overlay with unique images designed with paint, titling and animation programs. Since most of the time people can instantly see the difference between a computer generated image and real life you don't want to try and hide that fact. But in some situations you can trick viewers into believing that a computer generated image is something real.

The first way to do this is by simulating computer graphics. I know that sounds absurd but if you know anything about computers then you know that 99% of the people out there have no idea what computer graphics look like. Fortunately the Amiga can simulate the kinds of images that people expect. There have been a number of films and programs that have tried to simulate computer vision by putting graduated cross-hairs over a scene. The old television program The 6 Million Dollar Man made use of this technique even though there was no computer, no cybernetic eyes and the entire show was done on film. By overlaying numbers, lines, grids, etc., you can simulate computerized targeting, submarine periscopes, robot vision, etc.

The other way to trick viewers into believing that what they are seeing is real is don't keep it on screen very long. If you were outside taping a scene for a production and a UFO flew by, you might quickly swing your camera around, run after it, try to focus and then it would be gone. What would that look like when you played it back? Unless the UFO stayed nice and still while you lined things up you might get a handful of frames with a slightly out of focus, blurred, jittery disk. But those few frames would look more real than if you overlayed a nice, sharp, well drawn UFO in the center of the screen. If you want to overlay a UFO, have someone yell and then throw a Frisbee® behind you while taping. Try spinning around and shooting. Back in your studio analyze the tape and then try to recreate it with the computer and genlocking unit. You will probably find that the image is only visible for an instant or two.

You can also create ghosting effects by only partially overlaying graphics (not bringing up the computer or external video to 100%). By using a background color that is a tight 'weave' of color zero and another color you can even have a 'live' video ghost mixed with a computer background. One of the problems with genlocking devices is that you can overlay graphics on top of video but you can't overlay video on top of graphics.

## Fades

If you don't have fade controls on your camcorder you can use the genlocking unit to do this just by fading to a blank screen (not color zero). You can also fade into a scene using this technique. An advantage to doing fades with the genlocking device is that you can fade to and from any color (or pattern) and you can usually control the fade rate more precisely.

## Dissolves

Using the technique outlined in Chapter 5 and a genlocking unit you should be able to simulate a dissolve (where one scene changes into another with a smooth transition). You can, of course, dissolve from a computer image to a video image and vice versa with a genlocking device.

| | |
|---|---|
| **Animation Backgrounds** | With careful use of color zero you can use video for your animation backgrounds (rather than drawing them yourself). Just don't use color zero for your animated objects. |
| **Mis-adjustment** | Some genlocking devices have additional controls for adjusting saturation, chrominance and other levels. By setting these levels to extremes you can sometimes get strange effects. |

# Digitizer F/X

Digitizers have been used to generate special effects in video for the past few years. On the computer side, digitizers were the only way to get detailed visual information into the computer without drawing them by hand. In a strange way, video people appreciate the fact that digitizers look like computer images, while computer people have been trying to get digitizers to the point where they look like video images. Computer artists have made use of the fact that a digitizer is close to a photograph but just coarse enough so that it appears the image was created by hand. A crude, digitized image of Einstein can appear to be a very skillful computer artist's 'drawing'. You can make use of this if you need a painting as a prop, when you can't paint. With a program like Deluxe Photo Lab (see Chapter 6) you can blow up a digitized image and print it out in poster size (if you have a printer). With a touch up or two the poster will look like a skilled painting.

| | |
|---|---|
| **Print to Video** | There will be times when it is easier to use a digitizer to get an image from print to video. With the digitizer you can take printed graphics, bring them into the computer, clean them up or modify them and then transfer the images to video tape. This is particularly useful for industrial videos where you might want to include images from reports, charts, graphs, publicity photos, etc. |
| **Titling** | A digitizer is particularly good for creating custom title screens. About the easiest way to get a company logo, sketch, drawing, painting or any image into the computer so that it can be manipulated for a title screen is with the digitizer. Many digitizers have a black and white 'line art' mode that improves the sharpness of the digitized image. Many industrial video producers use the digitizer for this particular kind of custom title creation. It is much easier than trying to re-create a company logo with a paint or rendering package. If there is a particular font that you need that is just not available for the computer you can digitize the letters, put them into a paint program, and by making each one a brush you can create your custom titles. |

| | |
|---|---|
| **Black and White** | This is one of the easiest special effects you can do with a digitizer. Just turn the color controls down when you digitize a color image and you have instant black and white. |
| **Old Style Photography** | A variation on the 'color to black and white' effect that you can easily do with a digitizer is to make an image appear to be an old style photograph, tintype or Daguerrotype. Rather than turning the colors completely off to produce a black and white digitized photograph adjust the colors so that you have a brown and white, or sepia tone. With a few props you can make anyone appear to be a photograph of their own grandfather. For even more realism, pose your subjects in front of a video camera and go through the digitization process. The original photographs took minutes to expose so people had to remain still while their picture was being taken giving them a rather strange but unique look. You will get this same 'stiff' look with a digitizer. |
| **Newspaper Photography** | Another effect can be produced by combining the old style and black and white techniques to simulate newspaper photographs. Use a lower resolution setting than normal and reduce the colors to create a black and white or grey scale image. The pixelization is very similar to the old newspaper halftone process. |
| **Pixelization** | The thing that computer people try to avoid is the same thing that video people pay money for when it comes to digitizing. Pixelization is simply low resolution digitizing. Rather than setting the digitizer to it's finest resolution set it to something lower. The result is a much cruder image broken up into boxes. To get it even more pronounced use fewer colors. Pixelization effects can be used for titles or for simulating alien vision (or computer vision, which it really is). |
| **Mis-adjustment** | Your first attempts at digitizing will show you what happens when your adjustments are out of whack. You can create thousands of strange effects by telling the software that you are using the green filter when you are really using the blue filter, by moving the image while you are digitizing, by adjusting the hue, saturation and color levels, etc. If you want a photo of Mars just digitize any desert shot (or a sand box close up) with the red levels turned up. With a little fine tuning you can turn snow into sand, clear sky into blazing sunsets (or distant fires), and you can even get polarization, infrared or ultraviolet effects this way. Most of these kinds of effects you will find by experimentation. |
| **Multiple Images** | Since you have three 'exposures' during the digitizing process (red, green and blue) you can add and take away images during the process for a strange form of multiple exposure photography. This is a good way to put blue ghosts, green UFOs or pink elephants into any photograph. |
| **Montage** | By using transitions from one static image to another you can create a montage effect using a digitizer (or frame grabber). Just digitize (or |

grab) a series of images then use a presentation program or page flipping program to create a show to dissolve, cut (or use any transition effect that the software supports) from one image to the next. If there aren't too many images and the transitions are slow enough you can use a genlocking device to do this too. Just grab every other image then go back and forth from a video camera to the grabbed images. Put the camera on the first image, dissolve to the digitized second image, put the third image in front of the camera and dissolve back, bring up the next digitized image and dissolve back, etc.

**DVE**  DVE (Digital Video Effects) can be seen in just about every newscast these days. These effects are when a full size video image is expanded or shrunk and moved around the screen. Sometimes these special types of video effects are called ADO effects. An ADO (Ampex Digital Optical) is a special electronic device that can take a video image and shrink it, move it, distort it, flip it, etc. Like Xerox, ADO has come to stand for the effects it creates more than the device that creates them. Once an image is digitized it can be manipulated the same way that an ADO device manipulates video. But even during the digitizing process you can sometimes specify screen size, placement, etc.

## Frame Grabber F/X

Most of the frame grabber and frame buffer effects will be similar to those you can create with a digitizer (except for the multiple images effects, because the frame grabber works in much less time).

**Freeze Frame**  The simplest effect you can create using a frame grabber is the freeze frame. Just grab a frame then use the genlocking device to cut to the grabbed image during the transfer to the master VCR.

**Pixelization**  Since most frame grabber software gives you options for grabbing with different resolutions and a number of colors you can set the resolution low and reduce the number of colors to pixelize an image.

**Black and White**  Just set the color options to black and white or grey scale the same as you would with a digitizer.

**Old Style Photographs**  Same as with a digitizer.

**Newspaper Photographs**  Same as with a digitizer.

**Montage**  Again, you can use a frame grabber the same way that you use a digitizer to create a montage of stationary images. The advantage to using a frame grabber to do this is that you can take frames from moving video. Scenes from the rest of your production (people playing, working, etc.) can be put together into a title or credits sequence.

**Dissolves** As I outlined in Chapter 5 you can simulate dissolves with a frame grabber. End the scene on a static image, grab it, then using the genlocking device tape a dissolve from the grabbed image into the beginning of the next scene. When you edit the two scenes together they will appear to dissolve into each other.

**Mis-adjustment** Many frame grabbers let you adjust the hue, chrominance, etc., and by setting these up in the 'wrong' way you can create strange images. You can also try grabbing images from a VCR running at different speeds, or a camcorder that is mis-adjusted.

**DVE** Similar to the digitizer, you can sometimes set the frame grabber to reduce the size of the image that you are grabbing. This can be used to place a reduced size image in the corner of a screen (using a genlocking device) for weather maps, news stories, industrial videos, titling or credits.

## Paint Program F/X

You will have the most freedom in the area of special effects with a paint program. Used with genlocking units, image processors, digitizers, animation packages or just by itself you can create thousands of special effects with a paint program. Most of the effects done with a paint program are just a matter of creativity and a bit of paint program skills.

**Charts and Graphs**
Paint programs are ideal for creating attractive, professional looking charts and graphs. Later, you can even animate them using other programs.

**Touch Ups** An image grabbed or digitized can be brought into a paint program and cleaned up or modified. You can use the paint program to 'air brush' out unwanted elements (that beer can in the middle of a pristine landscape) or put in new elements (how would uncle Arnie look with a moustache?). A variation of this technique has been used in Hollywood for years. Movies would be made in the back lots of a film studio and later, artists would go back and paint out the telephone poles, wires, unwanted buildings, etc. A simple way to remove objects from a scene is to fill a paint program screen with color zero then genlock the video and the paint program. Now, select a brush and any color other than color zero. Bring up the brush color palette and try to make your brush match the color (or pattern) of an area right next to the object you want to remove. As you paint over the offending item it should disappear.

**Extra Objects**  As mentioned earlier, you could use a paint program to create almost any object (real or imaginary) and superimpose that object onto other video. Flying saucers, a full moon in an empty sky, balloons, statues, and on and on. All of these things can be created with a paint program and a little skill. All you have to do is make everything else in the picture (where you want the video to come through) color zero.

**Matte Painting**  Hollywood film makers have been using a special effect called matte painting for years. The film method for creating dramatic scenery was a very complex double exposure technique. The camera would be set up in one position for a scene to be shot. A sheet of glass would be placed in front of the camera and a line would be drawn around the objects in the frame to be saved (for example the horizon line). The glass would be removed and two masks would be made that exactly followed the outline, one for the area to be left alone and one for the area to be created or the matte. They would place the matte mask in front of the camera and the scene would be filmed with the unwanted areas masked out. This way only the action to be kept would be exposed on the film. Artists would then paint in the desired background on the glass (In the movie Ben Hur the hills of Rome were just a matte painting masking out the Hollywood hills. In the Wizard of Oz the emerald palace and the wicked witch's castle were both matte paintings). Back in the studio the film would be rewound, the other mask set up in front of the glass (to keep the original material from being exposed again), and the scene's second exposure would be shot. This time only the matte painting would be exposed on the film. When the film was developed they had the foreground they wanted (complete with moving actors) and the painted background they wanted.

This was (and still is) a very complicated, expensive and time consuming technique that you can simulate with a paint program and genlocking unit. Since the genlocking device will cleanly superimpose computer images on top of video you don't need sheets of glass, masks or double exposures (although a frame grabber will help quite a bit). Just shoot your scene the way you want, being careful that you don't move the camera and that no motion takes place in the part of the frame that you want to change later. Back in the editing studio grab a frame from the scene (if you don't have a frame grabber then you can genlock the scene and freeze frame the video or just play it back over and over). Now, using the paint program draw an outline completely around the parts you want to keep or the parts you want to change using color zero. Now you can turn off the VCR and concentrate on the matte painting. Fill in the parts of the screen where you want the video to come through with color zero. Everything else can be painted however you want. When you get the background the way you want it, go back and genlock the graphics over the scene during taping to the master VCR. If you were careful and creative enough with your painting you should have both the action video and the new background on the same screen.

With a digitizer or frame grabber you can take this one step further. Draw your outline on the video image as before, then make the outline into a brush. Using only the keyboard commands (leaving the mouse absolutely still) load in a digitized or grabbed IFF picture that doesn't use color zero (you may have to use an image processing package to adjust the picture ahead of time). Click down a copy of the outline on the new image (be careful not to move the mouse as the line may shift). Now, use the fill function to paint color zero over the parts where you want the video to come through. Genlock the picture during transfer to the master and now you have a digitized or grabbed image for your matte painting along with the untouched video in the foreground. While it is possible to do this with every frame of a video both of these methods work best with a stationary camera.

This may seem like a lot of trouble but the genlocking devices available today will only put graphics on top of video rather than video on top of graphics.

**Pointers**

Brushes can make quick and interesting pointers. When superimposed on video you can quickly and effectively draw the viewers attention where you want it. Obviously this isn't going to be appropriate for all video productions but for industrial and educational videos this method is a lot easier than creating custom animated pointers. Simply draw a pointer without using color zero (unless there are parts of the pointer that you want the video to come through). It can be as simple as an arrow, as complicated as a hand or something with special meaning to your viewers like a company logo in an arrow, a hot dog, a space ship or whatever. Turn it into a brush and then save it. You might wish to make copies of the pointer pointing in all directions. Now fill the screen with color zero, genlock your video and using the mouse to control the pointer you can move it around the screen manually. If your paint program has pages you can turn the pointer on and off by flipping pages or you can simply use the genlocking device. If you want a flashing pointer you can draw it in place then flash it on and off using the Undo feature.

**Color Cycling**

Color cycling can be used for a variety of effects from running water to falling snow. Check your paint program manual for tips on how to use color cycling. (Deluxe Paint III even gives you samples.) You can do simple animations with color cycling alone, if you want to do complex animations you should buy an animation program (see Chapter 7).

Anything that doesn't have too many colors or has only small repetitive movements can be done with color cycling. The shimmer of a Star Trek transporter, sparkling magic items, radioactive glows, electric short circuits, lightning, moving signs and dizzying op-art can be done with color cycling.

**Masks**  Quick and easy binocular effects can be created by drawing two overlapping circles filled with color zero, leaving the rest of the area black. Genlock the two and it looks like the view through binoculars. You can create any sort of mask this way. The view through a cat's eye, through a knot-hole, through a telescope, through venetian blinds, etc.

**Filters**  You can simulate various filter effects by creating a pattern that uses color zero and then partially mixing the image with video.

**Color Misadjustment**  By filling a screen with a color or colors and then partially mixing it with a genlocking device you can alter the overall colors of a video image. If you have a digitized or frame grabbed image you can adjust (or throw off) the colors by changing the color palette.

**DVE**  Once an image is captured (with a digitizer or frame grabber) you can make parts or all of the image into a brush and then using the brush manipulation tools found in most paint programs you can re-size, stretch, flip and distort the brush.

**Mapping**  Some paint programs give you mapping options where you can put an IFF image onto a shape. Imagine a photograph printed on a sheet of rubber. If you stretched the sheet over a ball you would be doing mapping. Computers do this digitally with a lot of fancy mathematics but the results are the same.

**Distortion**  Most paint programs include many features for drawing, creating, distorting, changing and manipulating images. Once an image is digitized or grabbed there are a thousand things you can do to it. When you mix these effects with video you can get some strange and interesting effects. Obviously, you may have more use for these kinds of effects if you are doing science-fiction, music videos, fantasy or horror videos but there may be instances when you might find something useful for industrial or educational videos. Watch an hour or two of Sesame Street and you will see a number of special effects that could have been produced on an Amiga with little more than a paint program and a genlock.

# Image Processor F/X

Image processing programs usually offer dozens of options and ways that you can manipulate graphics. Once an image is in an IFF format you should be able to manipulate it in many normal and abnormal ways. The manuals that come with these programs usually will give you a number of ideas and you will run across strange things on your own with a little experimentation.

**Image Enhancement** — One of the things that image processing software can do is enhance an image that isn't perfect. While you may not be able to turn an out of focus image into something that is crystal clear or add information that isn't in the image to begin with you can make most images a little clearer, cleaner, sharper, etc.

**Black and White** — Image processors are good for converting colors, averaging, reducing and enhancing images. With only a few key-strokes you should be able to convert an image to black and white, brown and white, blue and white or any other combination of colors.

**Color Mis-adjustment** — Since it is so easy to manipulate colors with one of these programs it is no problem to introduce false colors or change the colors to anything you want for a special effect.

**DVE** — Some image processing programs will let you simulate just about any DVE effect with an IFF image including, sizing, rotating, flipping, zooming and dozens of other digital manipulations.

**Mapping** — One of the most interesting special effects seen occasionally on MTV and in other places is mapping a video image onto a shape. Some image enhancers offer a feature that will do this. That way you can put your face on a soup can, a sphere or a number of other shapes. Some of the more sophisticated mapping techniques will let you use any IFF image to map onto any shape, even text.

## Rendering F/X

Rendering and ray tracing software is a step up (and slightly to one side) from paint programs. The objects and scenes you create can have a very realistic three dimensional look to them and yet will still appear computer generated. Until you get into very fancy ray traced images you will have a hard time disguising the fact that a computer drew something. But there will be times when you want people to know that (or at least wonder how) a computer drew something.

**Extra Objects**  Rendering programs are great for drawing three dimensional objects that can be rotated, flipped, distorted, manipulated and viewed from any angle. However, the objects created are rather blocky. You can use this 'blocky-ness' to create simulated computer displays of helicopters (or space ships) flying through cities, analyzing machinery or just about anything that can be drawn with a pencil and straight edge. They can also be used to create animations, logos or special text for titling. Because of their mechanical looking results they are not very good at drawing natural objects like human faces but they can produce realistic looking 'artificial' objects like airplanes, cars, etc.

**Re-Drawing**  An interesting effect that is a by product of the way that most structured drawing packages function is in the re-drawing. Every time you add another piece to one of these programs the entire screen has to be re-drawn from the beginning (some software will only re-draw the entire screen when you tell it to and other packages will hide the re-drawing process). If you create your entire drawing, begin taping and then tell the program to re-draw (or add another line to force it to re-draw) the result is a short scene of a graphic being drawn at remarkable speed. Granted, there won't be many times when this is appropriate, but it can make a straight forward display transition a little more interesting.

**Animation**  One of the more famous rendering animations was in the Dire Straights 'I Want My M-TV' video. The cartoon characters moving furniture were created with a rendering program. While it can be time consuming, an animation created with a rendering or ray tracing program can be quite effective. What you end up with (if everything goes right) is what most people used to call computer animation (naturally enough). To the viewer it is obviously not real, but at the same time it is obviously not Bugs Bunny type animation either.

One of the nicest animation effects that you can do with a rendering program is simulating smooth camera movements. Without changing the object at all you can have the 'camera' fly around, over, even through objects and scenes. Many rendering programs will even create the camera moving animation automatically. All you have to do is create the scene and then plot the movements, angles and speed of the 'camera', the program then renders all the frames for you and stores them in an ANIM file. Even though you may not be creating a lot of new objects this process can take a very long time (sometimes literally days of calculations).

# Animation F/X

The computer animation programs discussed in Chapter 7 will give you the more traditional, Walt Disney style animations. The characters, objects and backgrounds are more two-dimensional. Creating a sense of depth is entirely up to the artist. It used to take dozens of artists months or even years to create an animation but with the computer and the right software you can create cartoons in a few hours. Getting those cartoons to look like the professional ones will still take anyone a fair amount of time and talent but it can be done. There are hundreds of tricks developed over the years in getting animations to work smoothly. There are over two dozen books on traditional animation you can find (if you search hard enough). Unfortunately few of them deal with computer animation. If you are serious about creating lots of computer animations then you will still benefit from these books.

**Straight Animation**

With animation you can do just about anything, tell any story, use any actors in any setting or situation. There are almost no boundaries to what you can do. About the only limitation is that your viewers will know that it is just an animation (except for your very young viewers). Since cartoons have been a part of our lives for so long it only takes a moment for even the most sceptical viewer to reach the 'suspension of dis-belief' point, where they know it isn't real but they are willing to go along with it anyway. Most of the time, animations are just light entertainment but you can use them in almost any situation to get a point across or explain something (even government films about atomic bombs used cartoons to explain things). If you need to lighten-up a video or impart a piece of complicated information you might consider animation as a tool.

**Extra Objects**

Probably the most ambitious combination of real-life and animation is the film 'Who Framed Roger Rabbit?' This film is the ultimate example of adding extra, animated objects to scenes. If you have a few years, a lot of money and a lot of talent you could probably do your own version with a camcorder, genlocking device and the Amiga. On a much smaller scale you can still add your own animated pointers, titles, objects, even characters to your videos.

Unless you just want the animation to walk, bounce or fly through a scene oblivious to what is going on you will probably have to worry about positioning. In this case you will want to shoot your live sequences first. If your actors, or real elements will be interacting with the animation then careful planning is required. Using the same technique described in the matte painting section above, determine where the animation is going to be placed on the frame. You can then go back and create your animation using color zero as a background. When you are ready to put it in the live scene use the genlocking device to superimpose the animation on top of the video. Obviously, it is much more involved than this but those are the basic steps.

**Rotoscoping**   Almost as far back as the beginning of animation, studios have used a technique to simulate realistic character movement called rotoscoping. First, a real person would be filmed performing the desired action. The film was processed and then projected on a sheet of paper one frame at a time. Artists could then match their drawings to the real movements one frame at a time. The results were animations that moved as if they were real. Another variation of this was to project the original frames on a sheet of plastic or glass from behind, the artist would color or outline the projected frames and then expose a new film frame made up of both the original image and the artist's extra touches on top.

There are two ways to do this with the computer. First, you can grab every video frame (or every other frame), add your own touches and put them all together in an ANIM file for later transfer back to tape. This will take a frame grabber, paint program, a page flipping program, a great deal of memory and a lot of time. The second way to do this is by using a combination of the animation and matte techniques described above. (Hash Enterprises also sells an animation program called "Animation:Rotoscope, Traveling Matte Painting" that can simplify this procedure somewhat.)

# Titling F/X

You will probably use a titling program more than any other piece of software for putting together videos. For titles, credits, sub-titles or special labels a titling package is invaluable. But you can do more with a titling program than just simple titling.

**Straight Titling**

With a genlocking device you can put titles, credits or any words you want on top of video. Obviously, you can use overlayed text to create titles and credits but you can also use text to point things out and clarify things. Educational television makes extensive use of overlayed text and graphics to make things clear.

**Not So Straight Titling**

Without a lot of trouble you can create you own video version of the old silent films using titling software by simply cutting back and forth to the words. In the film Monty Python and the Holy Grail the director turned the opening titles into comic material (including sub-titles and sub-sub-titles). With a little imagination and the right sub-titles you should be able to turn the most boring piece of video into something humorous. You could even use a titling program to do video comic strips by putting speech and thought bubbles over your characters. Throw in a few "BIFF!"s, "BANGS!"s, "OOOFF!"s and "&*#%$"s and you are all set.

# Sound F/X

Sound and music effects can mean the difference between a professional video and just another home video. While you don't have to use the computer to do music or sound effects (and you probably won't be using the computer for voice overs) there are things that you can do to spice things up.

**Music**

It wouldn't be entertainment without music and it would be impossible to list all the effects that can be achieved with the right music. About all that I can say about music effects is that you should watch and listen to the great films and directors to see and hear how they use music to set moods. In many situations you don't even have to have a complete song. Many mysterious moods are set with nothing more than a few minor chords. In fact many times you will only have to hint at a type of music with a few chords and then fade the music down or out completely.

**Sounds**

The kinds of sounds that you might want to create with the aid of the computer are the ones that can only be produced electronically. For example the strange roars in Star Wars were recordings of lions played backwards. The computer is good at taking sounds and shifting them up or down in pitch, volume, playback speed, etc. You can also take a single sound and multiply it over and over (creating an entire audience out of one person clapping). Most of these effects will require an audio digitizer to get the sound into the computer in the first place but there are disks filled with nothing but sampled sounds that you can buy.

**Foley Sounds**  Foley artists were the unsung sound effects people of radio days. They would have sheets of glass for breaking, creaky doors, boots, horns, bells, whistles and thousands of other exotic and ordinary items for creating the 'normal' sounds people expected to hear. Films and television programs also have foley people to supply all the extra noises, bumps, steps, etc. You might want (or need) these extra noises in a video production. You should spend a lot of time trying to eliminate extraneous noises during taping but in the process you may be eliminating normal sounds like footsteps. With a little imagination and a few items found around the house you should be able to put some of those sounds back into your videos. The computer can help mainly in the area of timing or altering the pitch of a sound (the same footstep sound could be altered to sound like a man, a child or an army walking on a wooden floor, a tile floor, dirt or just about any surface. The computer can also add echo, make a sound deeper, rounder, more resonant or completely distort a sound so that it is unrecognizable.

Of course you may want to try eliminating some sounds with the mixer or recording. For example, a car horn momentarily honks in the background but no one is speaking. You could try and take that sound out. With the computer you can identify and sometimes eliminate sounds by smoothing out a digitized wave.

The thing to keep in mind about creating those 'normal' sounds with digitizers and sound manipulating software is that it is much easier to create strange and unusual sounds than to do specific things.

**Voice Overs**  You might not think of a voice over as a special effect but it can be an effective tool. From straight narration to humorous additions voice overs, like music, can enhance or completely change the visual. For example, let's say you have a plain shot of a ball on a table. Changing the voice over can completely change the way that people 'see' the ball. "This is an ordinary child's toy", "This is a bomb", "This is an alien space ship", "This is supper", "This is Mars", "This is the size of a neutron star", "This is my mother-in-law." You get the idea.

**Sampled Sounds**  Sound samplers or digitizers are great for getting audio into the computer. Once there you can play it back just the way it came in or you can modify it. Generally, sound digitizers are used for musical applications but you can also use them to 'record' any noises you want, including speech. Because the quality of the sampled sound is rarely as good as a straight recording you probably won't want to digitize all your extra audio but there are advantages sometimes. The biggest advantage to digitized sounds, apart from being able to modify them, is that once they are in the computer they can be linked with other computer generated events like animations and presentations. It might be easier to digitize dialogue, modify it a little and then link it to an animation than to try and lip-sync the voices. Another advantage to this is that you don't have to be a master of different voices to create multiple characters. By speeding up, slowing down, changing the pitch or fooling with the software settings you can create hundreds of different voices from the same sample.

**Created Sounds**  Sampled sounds are also good for creating those special sound effects like laser guns, space ships, monsters, etc. By recording different things and then playing with the adjustments you can create thousands of unique sounds. Some music programs will let you make sound adjustments to the instruments included. Try changing the settings a little or a lot and then playing them back as a single note, you might find just what you want. This process is pretty much hit-or-miss so if you run across a sound that you think you might want later save it and make a note.

## Summary

Obviously, I have only scratched the surface of special effects. Pick any one of these ideas and you will be able to come up with a dozen variations. I'm sure that I have even left out entire categories (like makeup, special props, animals, explosives, etc.). Try your library for more information about these areas.

Again, it is best to use special effects sparingly (or they aren't that special any more). You will have to experiment to get special effects the way you want them. You should also try to keep in mind that the computer is very good at some things and not so good at others. Realistic special effects take a lot of planning, time, experimentation and the most important element imagination.

# Chapter 11

## Putting it all together

# Chapter 11
# Putting it All Together

I've talked about all of the elements of desktop video and you probably already have a good idea about how all the parts work independently. Now it is time to put it all together into a complete package. In this chapter I'll talk about a few different configurations, realistic scenarios and give you some ideas. You should be able to use these as outlines for productions of your own.

First, I'll talk about some basic system configurations and then talk about extras you will need for specific applications. Finally, I'll talk about planning and producing some videos.

## System Configurations

There are many different combinations of software and equipment you could have all depending on what you need to do and your budget. If you are going to be doing nothing but titling then your requirements will be one thing, if you want to experiment with lots of different desktop video ideas your requirements will be something else. If you are planning to use desktop video for your own amusement or if you are planning to use desktop video in a professional application you will need different things.

What you will want to buy also depends on what you already have. If you are more of a computer person then you may need to look into getting the right video equipment. If you are already doing video productions then the computer side is where you will have to focus your attention. Here are a few ideas on basic desktop video setups in different price ranges.

The first step in putting together a system is planning what you need now and what you expect to need later on. Some of the things you buy in the beginning can be sold later on but most of the items will be yours forever. It is going to be hard, psychologically, to go out and buy a $1300 genlocking device three months after you spent $750 on a different one. Some of the things you buy can be upgraded and some of them might just end up in the closet.

The first consideration is going to be where your productions will be seen. If you are mainly interested in home videos and have no plans to ever try and get them on the air then you should be able to get by with something less than broadcast quality equipment. If you think (or know) that your productions will be used by other people (for example; if you are thinking about just doing title screens, graphics or short animations for companies that might include them in their own productions) then you should get as close to broadcast quality as possible. If you think that your productions will be going through a number of generations (say for educational purposes) then again you should be thinking about near broadcast quality. If you are thinking that eventually your work will be shown on cable or even broadcast television then definitely look at the high end equipment.

The same applies on the software side. If you are mainly interested in animation then you will want to focus on those packages. If all that you want to do is titling then those are the software programs you should be spending your money on.

In all of the following lists I have tried to use the suggested retail prices but you will probably be able to get better deals in most cases. Buying anything through the mail (particularly expensive hardware) can be a little risky. But if you are careful, buy through established mail order houses, check on the return policy, make sure that the items you want really are in stock and not on order, and use a credit card or COD then you can save a lot of money on software and smaller items. In any case you should shop around for the best buys.

## Minimum Configuration

If you are just getting started and are on a tight budget you are going to have to be very careful about what you buy and when. You should also think about what you plan to do later on. If you want to get started right away but also think that you will eventually be doing more serious work then it might be wise to think about some of the more expensive items right from the start. The items in this list are mainly for home applications. Here is a list of the items that you will probably want for a minimum configuration.

Camcorder - ($600 to $1500) Almost any camcorder will do for a minimum home configuration. If you are going out to buy your first camcorder you should first buy a handful of video magazines (a list appears in the back of this book) and read everything you can about them. You should also go back and read Chapter 2 if you haven't already. I would stick to either 8mm or VHS because they are an accepted format, you have a wide choice, and they work pretty well. The more important features to look for are flying erase heads (very handy but not absolutely essential), fade controls, external control ports, dubbing features (including video and audio input ports), fairly low lux number, macro lens and at least a 6 to 1 zoom. You can get by with any, all or none of these features but they will come in handy later on.

VCR - ($250 to $500) Once again, it is probably a good idea to read as much as you can about VCRs before you go out and buy one. For home applications you will probably want to stick to VHS just because you can share your productions with friends, there are dozens of models to pick from, and you can rent movies. The odds are that you already own a VCR but if you don't or you are planning to buy a second VCR for your desktop video studio then here are a few things to look for. Flying erase heads (either your VCR or camcorder must have flying erase heads to do any desktop video work), audio dub capabilities (very handy for eliminating extra generations), HQ circuitry, linear time counter (not precise but much better than tape counters), external control ports (for connecting to an editor/controller), four or more heads, single frame advance and slow motion. About the only one of these features that is absolutely necessary are flying erase heads. If your camcorder has flying erase heads and you are planning to use it as the master unit then your VCR should be the same format. That way you can shoot with the camcorder then play the same tape back in the slave VCR. Otherwise, you will have to make a dub to the VCR's format before you even begin editing.

Amiga 500 - ($750 to $1000) The reason that I give a range for the A500 and not a specific price is that when you buy the computer the dealer will often sell it as a package, including the monitor, extra memory, sometimes an external disk drive and some software. It is usually a good idea to get the complete package because it is cheaper than buying the components separately and you will be assured that all the parts will work together. The Amiga monitor is very good for the price and anyway, you will need it eventually. Definitely get the A501 internal memory upgrade board. While you can get by with the standard 512K that comes with the A500 you will soon need to expand it to at least 1mb (one mega-byte). You will also find that a second disk drive, while not critical, will save you a lot of disk swapping.

Encoder - ($80) This really is an absolute minimum configuration device but if you can't afford anything else then the Creative Microsystems VI-500 will at least let you record computer images on video tape. You won't be able to superimpose graphics on video or many of the other things that I talked about before but you will be able to do titles, graphics and animations (with the right software that is). If you think that you will eventually want to do any of the fancier things then it would be better to wait and get a genlocking device with a built-in encoder. I mention it here because it is an option that will at least get you started.

Paint Program - ($100 to $150) If you don't buy any other software for desktop video you should buy a paint program. Most people buy Deluxe Paint from Electronic Arts but any of the paint programs I talked about in Chapter 6 should work fine. With a paint program you should be able to do some titling, some animation, and of course, some graphics. Paint programs won't do as good a job at titling or animation as the dedicated programs but with a little extra work you should be able to produce reasonable results. If you are just buying a new A500 see if you can get a paint package thrown in for free, or at a discount.

Miscellaneous Items - ($100 to $300) No matter what your budget you will have to buy a number of extras. For the camcorder you will need blank tapes, some extra lights, extra camcorder batteries, perhaps a tripod and something to carry it all in. For the VCR you need more blank tapes, and sooner or later a head cleaning tape (I don't recommend cleaning the heads until you absolutely have to because any cleaning puts extra wear on the heads. Also you should get a non-abrasive head cleaning tape. If your heads get clogged to the point where one of these won't help then take your VCR in to get it cleaned professionally). On the computer side you will definitely need some blank diskettes. To hook everything together you will need cables. You will find that the miscellaneous category will continue to grow as you do more.

Adding it all up so far we have spent the following (I'll just use averages here):

| | |
|---|---|
| $ 1000 | Camcorder |
| $ 350 | VCR |
| $ 1000 | Amiga 500 (with monitor and extra memory) |
| $ 80 | Encoder |
| $ 100 | Paint Program |
| $ 200 | Miscellaneous Items |
| $ 2730 | Total |

This is just about the absolute minimum configuration for doing any kind of desktop video. You probably already have a few of these items so let's look at improving and adding a few more items. Keep in mind that we are still looking at a basic system, probably not good enough for industrial, educational or broadcast use but definitely good enough for home use.

Genlocking Device - ($450) Rather than an encoder you might think about getting a genlocking device instead. If you are going to get a genlock (and you probably will) it is better to buy a good unit from the start. I would suggest the ProGEN from Progressive Peripherals is the best for under $500. There are other, less expensive units available, but I think you may be disappointed with them. They will work but for an extra hundred or so you can get a much better unit from the beginning.

Digitizer - ($280 to $480) DigiView Gold from NewTek is a great little device that you should be able to think of a thousand uses for. There are two options when you buy this digitizer. You can either buy a black and white video camera to use with it (about $280) or buy a color splitter (about $80) and use your camcorder. Using a black and white video camera will give you the best results but there is the extra expense and you end up with another camera that isn't good for too much else. The Panasonic WV1410 recommended by NewTek isn't a bad black and white video camera but it is basically a security camera that is not portable). The color splitter won't give you quite the resolution of a black and white camera but you save a lot of money. With either of these methods you will need some extra lights and a tripod (or improvised camera stand).

The next few steps are a matter of what you want to do with your videos. If you just want to do titling (straight text only) and nothing else then you don't need a digitizer or a paint program (but I strongly recommend a paint program anyway). If you are interested in simple animations then an all-in-one animation package might be the next step. If you want to add music and voice overs then you will need a mixer, a microphone and a music program. Let's look at some of these.

Animation Program - ($60 to $140) The all-in-one animation programs are fairly easy to use and you can do a lot with them. Of course you are limited to the two dimensional cartoon style animations but that may be all that you need. They can be a lot of fun and even if it would be difficult to do Disney quality you can still impress your friends and family.

Mixer - ($40 to $100) To do any sort of music or voice overs you will need a mixer of some sort. Radio Shack has two or three, inexpensive mixers that should be good enough for desktop video work. While you can get by with only two or three inputs you will save yourself a lot of cable switching with more inputs.

Sound System - ($150 up) To go with recording, playing back and listening to your audio you should have a sound system. You can use your average home stereo for all of this if you want to or you can go out and buy one of those cheap all-in-one units that has a turntable, amplifier, speakers, cassette deck, etc. About the only requirement is that you should have a cassette deck. It doesn't have to be top-of-the-line equipment, just good enough to make a fairly clean recording. Anything above a walkman/portable/boom-box type system should work fine as long as the cassette deck has auxiliary input and output ports so you can connect it to the mixer.

Microphone - ($20 to $50) Just about any reasonable electric condenser microphone should work for voice overs. Again you might check Radio Shack for one (you can make a stand yourself).

Music Program - ($20 to $100) If you just want an idiot-proof music program for quick background pieces then you can get DNA Music, Fractal Music or Protein Music (all $20) from Silver Software or Instant Music ($50) from Electronic Arts. If you are more interested in composing your own music then Deluxe Music Construction Set ($100) from Electronic Arts is both a good program and not too expensive.

Titling Program - ($80 to $100) While you could use the paint program for titling you will get better results with a titling program. Two effective and inexpensive titling packages are Animation:Titler ($80) from Hash Enterprises and TV*TEXT ($100) from Zuma Group. Titler, while it does use the mouse and pull down menus, will take a little work to use. TV*TEXT will help you put together very nice title screens but has no transition capabilities.

Image Processing Program - ($40 - $150) Just to make sure that these programs work together and with your genlocking device you might want an image processing program. Butcher ($37) from Eagle Tree.

So let's see where we are again.

| | |
|---|---|
| $ 1000 | Camcorder |
| $ 350 | VCR |
| $ 1000 | Amiga 500 (with monitor and 1mb) |
| $ 450 | Genlock (ProGen) |
| $ 280 | Digitizer (with color splitter) |
| $ 150 | Sound System |
| $ 60 | Mixer |
| $ 30 | Microphone |
| $ 100 | Paint Program (DigiPaint) |
| $ 20 | Music Program (DNA, Fractal or Protein Music) |
| $ 100 | Animation Program (Fantavision) |
| $ 80 | Titling Program (Animation:Titler) |
| $ 40 | Image Processor (Butcher) |
| $ 200 | Misc. |
| $ 3860 | Total |

If you had to go out and buy all of this at once then it might seem like a lot of money for a home desktop video setup. Keep in mind that you probably already have a lot of these items and you can probably get better prices than these listed. For many of us $4000 is a lot of money but consider what you are getting! If you buy everything on this list you will have all the parts necessary to videotape, edit, digitize images, do voice overs, create music, do animations, create professional quality titles and credits, have some of the best computer graphics possible, overlay those graphics and titles, do thousands of special effects, plus you will have a VCR for watching rented movies and one of the most powerful home computers ever created. You would have to spend almost twice that much just for a new Macintosh II computer, a professional quality VCR or a low-end character generator.

## Mid-Range Configuration

There are no hard lines between minimum, mid-range and professional systems. In the case of software the differences will sometimes be small or exactly the same for all three. For the purposes of this book let's assume that mid-range applications are going to include the serious high-end hobbyist and go up through educational and some industrial applications. With the right camcorder and VCR you would be getting close to broadcast quality. Obviously, you can mix and match from all of these categories depending on what you want and need.

Camcorder - ($1300 to $2500) I am giving this kind of range for a mid-level camcorder to cover everything from a very good VHS or 8mm up to Hi8 and S-VHS. You will want to look at all the features described above for the lower-end camcorders (flying erase heads, dubbing, fade controls, etc.).

Lenses and Filters - ($20 to $200) As your productions get more sophisticated you might want to think about getting some special lenses and filters for your camcorder. The first lens you might want is a wide angle lens. Most of the lenses built into camcorders these days concentrate on extreme closeups or extreme zoom ratios but aren't very good at wide angle shots. The first filter you should get is a neutral density filter to protect your main lens. There are many different lenses and filters for creating special effects and for special needs.

Microphones ($40 to $250) As you move up you will also want to get some specialized microphones. Radio microphones and telephoto or zoom microphones will enhance your remote shoots quite a bit. For special camcorder mikes check the video magazines (Azden has a complete line of microphones for video use).

VCR(s) - ($500 to $1200) When we start getting into mid-range to high-end systems we can start looking into special features beyond the required flying erase heads. You will probably want to be looking at consumer editing decks with features like audio dub, jog-shuttle controls, assemble as well as insert editing, and overall more accurate editing capabilities. You will probably be ready to buy a second VCR too. This will save wear and tear on your camcorder and a master/slave two deck set-up is much more efficient. The odds are that you will be buying a mid-level VCR for the special editing features and using your old VCR as a slave (curiously enough the deck with the fancy editing features might be better for the slave unit because the master just has to stop, start and record at the right times while the slave unit is the one where you need the most accuracy in locating scenes, backing-up, jumping forward, etc.). If you are looking at S-VHS or Hi8 decks you might have to spend more than this for these special editing features.

Editor/Controller - ($200 to $750) There are two ways that you can go in terms of an editor/controller. Either buy a unit specifically designed for use with your master and slave VCRs or use the computer to control your decks. There are some companies listed in Appendix A that will modify your decks and have the proper software for doing this. There isn't much to look for in an editor/controller except to be sure that it will work with your equipment. Most of them perform pretty much the same functions and at the consumer level if the manufacturer even makes an editor/controller in the first place they only make one unit.

Image Enhancer/Color Corrector - ($50 to $1500) There are a number of video image and color correctors that range from simple filters or signal boosters all the way up to what amounts to a home level proc-amp (processing amplifier). While these units won't improve your picture beyond the limitations of your format they can help reduce the signal degradation going from one generation to the next, help improve a weak signal and can fine tune your colors during dubbing and editing.

Amiga 2000 or 2000HD ($1500 to $2500) Like buying an Amiga 500 you can usually get a good price if you buy a package deal from a store. You will probably want the monitor and a second disk drive. If you can afford it you might be better off with the 2000HD which includes a built-in 40 megabyte hard disk drive. The main reason for getting a hard disk is that when you start working with screens of information, graphics and animations you need a lot of storage space.

Memory Board - ($400 to $1200) The A2000 series computers all come with 1mb already but you might want to consider buying an extra memory board anyway. Memory boards for the A2000 usually come with either 0, 2, 4, 6 or 8 megabytes. Zero mb or unpopulated boards require that you buy the memory chips separately and plug them in yourself. Actually, most memory boards only come with 2mb installed. If you are uncomfortable installing chips and boards then you can usually get your dealer to do it for you (which is one of the reasons that you might want to buy your Amiga at a dealer in the first place).

Hard Disk Drive - ($500 to $1200) If you didn't buy an A2000HD or you want to upgrade your existing A2000 for more serious video work then you will probably want a hard disk drive and controller (the controller is just a board that connects the hard disk drive to the computer internally). There are many hard disks and controllers available for the Amiga and their prices and storage capacities vary quite a bit. Hard disks get more economical the larger the storage capacity. A ten megabyte drive may cost $500 but a twenty megabyte drive might cost $650. Once you start using your hard disk on a regular basis you will probably use up most of a 10mb drive fairly soon so you should probably start out with a 20mb or larger to begin with.

Genlocking Device ($750) The next step up in genlocking devices would probably be the SuperGEN from Digital Creations. It produces a better signal than the ProGEN and offers a few more professional features.

Frame Grabber - ($700) Progressive Peripherals makes a very nice frame grabber that will do just about everything that a digitizer will and more. With a frame grabber you can start increasing the flexibility of your system and add to the number of special effects.

Audio Digitizer - ($90 to $120) An audio digitizer with software will let you bring in sounds and voices which you can then manipulate and include in your music program and/or tie in with animations.

MIDI Interface - ($50 to $250) If you are getting into music composition software then you might want to think about MIDI devices. To connect the Amiga to a MIDI device you will need a MIDI interface. The MIDI interface itself is fairly simple to use but you should know something about music to get the most from the software and MIDI instruments.

MIDI Instrument - ($200 up) The choice of MIDI instrument is entirely up to you. What instrument you play or compose upon and your budget will determine what you buy. There are dozens of MIDI compatible instruments available these days ranging from drums to keyboards to professional synthesizers costing tens of thousands of dollars.

Paint Program - ($100 to $150) You still need a paint program and fortunately just about all of the ones mentioned in Chapter 6 are very good. Good enough, and inexpensive enough to fit in all three categories. They each have their own special features and most computer artists own more than one of them.

Rendering Program - ($150 to $500) Starting to get into the more serious graphics packages we have rendering and ray tracing programs. As I mentioned in Chapter 7 these programs produce stunning graphics but are more difficult to master.

Image Processing Program ($40 to $150) Once you begin to mix graphics from various programs the need for an image processing program increases. You will need to convert formats, resolutions and adjust color palettes.

Animation Program - ($100 to $150) You may still want an animation program for producing quick, two dimensional animations and storyboards. In fact an animation program may be all that you want depending on your needs.

Page Flipping Program - ($60 to $150) These programs will give you more control over animations and titling transitions.

Titling Program ($100 to $150) At this level you might think about splitting your titling jobs into a screen/text generating program for your titles and a presentation program for your transitions. If you will be doing a lot of productions then you might want all three (screen/text, presentation and all-in-one titling programs).

Presentation Program - ($70 to $150) Like a page flipping program you will have more control over your animations and titling.

Music Program - ($20 to $150) Again, you will want a music program. If you aren't musically inclined you can stick to the idiot proof programs and still produce decent background music. Or you can move up the the music composition programs with MIDI control.

Let's total it up.

| | |
|---|---|
| $ 2000 | Camcorder (Hi8 or S-VHS) |
| $ 150 | Lenses and Filters (wide angle lens and filter kit) |
| $ 250 | Microphones (stand, radio and zoom mike) |
| $ 1600 | 2 VCRs (an editing deck and a Hi8 or S-VHS deck) |
| $ 250 | Editor Controller |
| $ 250 | Image Enhancer/Color Corrector |
| $ 1600 | Amiga 2000 (with monitor and extra disk drive) |
| $ 400 | Memory Board (2 megabytes) |

| | |
|---|---|
| $   800 | Hard Disk Drive (40mb with controller) |
| $   750 | Genlock (SuperGEN) |
| $   700 | Frame Grabber (Progressive Peripherals) |
| $   200 | Sound System |
| $   120 | Mixer |
| $    90 | MIDI Interface (Midi Gold) |
| $   250 | MIDI Instrument (keyboard) |
| $   100 | Paint Program (Photon Paint) |
| $   150 | Rendering Program (PageRender3D) |
| $    50 | Image Processing Program (PIXmate) |
| $   100 | Animation Program (MovieSetter) |
| $   150 | Page Flipping Program (PageFlipper Plus F/X) |
| $   100 | Titling Program Screen/Text (TV*TEXT) |
| $   200 | Titling Program All-In-One (Pro Video) |
| $   100 | Presentation Program (TV*SHOW) |
| $   150 | Music Programs (Deluxe Music Construction Set and Instant Music) |
| $   200 | Misc. |
| $11610 | Total |

When you list it all like this it seems like quite a bit and it is. Not only are you getting just about everything you could want in a very nice home video system you are getting just about everything in a home computer system too. You could probably shave a few thousand dollars off this figure if you aren't interested in MIDI, multiple animation and titling programs, use equipment that you already have and buy at a discount. On the other hand you could easily go over this total if you buy top-of-the-line equipment. If it still seems like a lot of money try pricing a video paint box, a digitizing music synthesizer or a MAC II.

With this system you should be able to do just about anything in this book and more. You will be able to do educational and industrial videos (even some broadcast applications).

## Professional Configurations

If you are thinking about professional level desktop video productions then you probably know what kinds of camcorder and VCRs you want. Most of the hardware and software will be the same as in the mid-level configuration. Things begin to diverge at this level depending on what your application will be. The system I'll describe is more of a Christmas wish list, covering the ultimate in desktop video hardware and software. You could of course spend more money (particularly on the video side) and you probably won't need all the things on the computer side. But if you are serious about titling, animations, music and special effects then these are the kinds of things you will want.

The list below is about as far as you can go without entering into the broadcast area. You should be able to do a number of broadcast level things with the following equipment but if you are setting up a cable or broadcast studio then you will want more sophisticated equipment than this.

Camcorder - ($7000 up) About the minimum camcorder you will want is ED-Beta if you plan to have your productions close to broadcast quality.

Lenses and Filters ($200 up) You will want some nice, high quality lenses and a range of filters.

Microphones - ($500 up) An automatic zoom mike, radio mikes, stand mikes and lavalier mikes.

VCR - ($3000 up) Again an ED-Beta deck will cost you at least this much and you will want two of them.

Editor/Controller - ($1000 up) This is a guess on the price (and a low one at that).

Image Enhancer/Color Corrector - ($700 up) When you start getting to this point then one of these become less a luxury and more a necessity.

Waveform Monitor - ($1500 up) Another guess on the price. Again, when you get to this level then you should seriously consider getting this kind of signal testing equipment. This will help in avoiding illegal colors from the computer.

Vectorscope - ($1500 up) If you get a waveform monitor then you should get a vectorscope at the same time.

SMPTE Time Code Generator/Reader - ($500 up) A SMPTE (Society of Motion Picture and Television Engineers) time code device will give you frame accurate editing capabilities as well as a way to tie into MIDI devices.

Sync Generator - ($2000 up) If you want a nice clean signal then start with nice clean sync. Many sync generators also have test signal generators built in so you can adjust all your equipment. This may not be a critical piece of equipment but if we are wishing for the ultimate system lets get one.

**Switcher/Fader** - ($1800 up) For multi-camera studio work or fancier editing you might want a switcher/fader. These can get very, very expensive and not everyone will need one.

**Amiga 2000HD** - ($2500 to $3000) This is not really an outrageous amount of money for what you get. A top of the line Amiga will cost you less than a Macintosh. You should also get the monitor and an additional floppy disk drive. The A200HD comes with a 40mb hard disk drive built in.

**Memory Board** - ($800 to $1000) For extensive animations you will need as much memory as you can get which is 8mb on an Amiga. This may seem like a lot of memory but you would be surprised how fast you can eat that up.

**Accelerator Board** - ($1200 to $2000) If you want to do ray tracing animation than you will want a 68020 or 68030 CPU (Central Processing Unit) and a 68881 or 68882 math co-processor. This will speed up your rendering times significantly.

**Printer** - ($250 up) A printer is not really part of the desktop video set up but it can help you put together scripts, EDLs (edit decision lists), storyboards, letters, etc.

**Genlock** - ($1700) The top-of-the-line genlocking unit is the Magni 4004 used by most professionals. When you get up to this level and this price range you should be able to test and evaluate a genlocking unit yourself so you might consider the Neriki.

**Frame Buffer** - ($1000) This unit will perform all the functions of a digitizer or frame grabber, plus it will let you hold video frames using a wider range of colors and resolutions. You will also be able to render objects to the frame buffer for later display or processing.

**Sound System** - ($500 up) you can still get by with an inexpensive system but if you are serious about your music then you might consider getting a much better system. Again you need to be able record and play back cleanly.

**Audio Mixer** - ($120 up) You can spend thousands of dollars on a studio level audio mixer but you are still limited by the reproduction abilities of video systems (which aren't that great).

**MIDI Interface** - ($80 to $150) These are the same prices at the top or bottom of the scale.

**MIDI Instrument** - ($200 up) Again, what you get is up to what you play and what you can afford.

Music Program - ($50 to $300) From idiot-proof music creators to MIDI sequencing software it all depends on how deep you want to get.

Paint Program - ($100 to $150) The top and the bottom are still the same and you will still want a paint program. Deluxe Paint III is probably the best but you might want to try more than one program.

Rendering Program - ($500) The top of the line rendering and ray tracing program is probably Sculpt/Animate 4D but most animators also own Turbo Silver ($200) and use both programs. You might also want Interchange from Syndesis to move objects back and forth between these two programs.

Animation Program ($100 to $900) Even if you are spending a lot of time with the rendering programs you might think about using an all-in-one animation program to get a quick idea about how things will look and for storyboarding. If you are serious about animation then you might look at the entire line of programs and utilities from Hash Enterprises.

Page Flipping Program ($60 to $150) These will give you much more control over your final animation.

Image Processing Program - ($100 to $150) Again, you will need one of these programs when you start mixing different graphics packages.

Titling Program Screen/Text - ($70 to $200) You will want one of these to create custom title screens.

Titling Program Screen/Text/Display ($200 to $300) The two top all-in-one titling programs, in my opinion, are Broadcast Titler from InnoVision Technology and Pro Video Gold from Shereff Systems. Both are very easy to use and produce excellent results.

Presentation Programs ($100 to $200) For bringing those custom title screens to life you need a presentation program.

So let's total up our ideal system.

| | |
|---|---|
| $ 7000 | Camcorder (ED-Beta) |
| $ 200 | Lenses and Filters (wide angle lens and various filters) |
| $ 500 | Microphones (stand, zoom, radio and lavalier mikes) |
| $ 6000 | VCRs (two ED-Beta decks) |
| $ 1000 | Editor Controller |
| $ 700 | Image Enhancer/Color Corrector |
| $ 1500 | Waveform Monitor |
| $ 1500 | Vectorscope |
| $ 500 | SMPTE Time Code Generator/Reader |

| | | |
|---|---|---|
| $ 2000 | Sync Generator | |
| $ 1800 | Switcher/Fader | |
| $ 2500 | Amiga 2000HD (with monitor and extra drive) | |
| $ 800 | Memory Board (8 megabytes) | |
| $ 1200 | Accelerator Board (with math co-processor) | |
| $ 250 | Printer (dot matrix) | |
| $ 1700 | Genlock (Magni 4004) | |
| $ 1000 | Frame Buffer | |
| $ 500 | Sound System | |
| $ 120 | Audio Mixer | |
| $ 100 | MIDI Interface | |
| $ 200 | MIDI Instrument | |
| $ 150 | Music Program(s) | |
| $ 150 | Paint Program (Deluxe Paint III) | |
| $ 500 | Rendering Programs (Sculpt/Animate 4D and Turbo Silver) | |
| $ 500 | Animation Programs | |
| $ 150 | Page Flipping Program (Page Flipper Plus F/X) | |
| $ 150 | Image Processing Program (Deluxe Photo Lab) | |
| $ 200 | Titling Program Screen/Text (TV*TEXT Pro.) | |
| $ 300 | Titling Program (Broadcast Titler) | |
| $ 150 | Presentation Program (Deluxe Productions) | |
| $ 400 | Misc (even extras get more expensive at this level) | |
| $33720 Total | | |

Of course you might not want or need most of this equipment (almost $12,000 worth of special video test equipment, sync and SMPTE time code generators for example). But if you talk to some of the people doing serious desktop video work with near broadcast quality equipment you will find that most of them have spent more money than this on their set-ups. I would also challenge anyone to go out and buy any other kinds or combinations of equipment that could do all the things that this configuration could do.

# Step by step

Well, now that you have all your equipment (and nothing left in your bank account) you will want to do something with it all. That is where this section (and this entire book) leads to.

There are really three sections to putting together a video production; Pre-production, production and post-production. Each of these can be further broken down into smaller parts. If there were hard and fast rules about producing a film or video then we would all win Academy

Awards (or no one would) so these are just suggestions on how to approach the problems.

## Pre-production

You could say that everything that isn't actually shooting or editing is part of pre-production. Reading this book is a form of pre-production. Pre-production involves planning, scripting, preparation, practice and rehearsal. We can take a look at each of these.

Planning

Planning starts when someone says "I have a great idea for a video!" This could be you, your boss, a company or anyone. The trick here is to step back a little bit and start asking a few questions.

Why is this a great idea? The answer could be because someone else wants it, needs it or would enjoy it, or the answer could be just because you want to do it.

Who is going to see it? Clients, kids, friends, teachers, scientists, people at a trade show, company executives, prospective employees, etc. The point to asking this question is so that you have an idea how to begin the rest of your planning. If you are taping a wedding then you probably won't be using a lot of science-fiction effects or animations. If you are doing a production for kids to explain multiplication tables then you might need lots of graphics and animations. If you are doing an auto repair instructional video then you will need lots of extra lighting for the garage shots.

You might also ask yourself where are people going to be seeing this? In your living room, in someone else's living room, in a board room, in a classroom, in a theater, at a trade show, in a display window? In your living room or in a theater you can take more time developing your story because people have made an emotional investment in seeing your work. However, if you are using the video in a lecture situation where people will be seated far from the monitor you should avoid using small type fonts in titles, and stick to large, clear images. In a board room you want to get your point across quickly and efficiently without a lot of frills that might be considered time wasting. In a classroom your audience needs to be entertained as well as instructed or they will lose interest quickly. If your production will be shown at a trade show or in a store window you want lots of flash and glitter to grab people's attention with almost every shot. You also want to get your points across quickly and repeatedly because people will be coming in at the beginning, middle, end and all points in between. It is better to be sketchy than risk losing your audience. On the other hand, if you are doing an instructional video then people will want you to be very detailed and very clear.

How long will the production be? The answers will determine the style you choose. If you are only going to have 30 seconds for a commercial then you have to budget your shots carefully. If you are doing a program for broadcast then you might have to leave time for commercial breaks and keep to a very specific time frame. If you have all the time in the world then think about balance, continuity and the fact that few people will sit still for a four hour epic unless it is very special.

How long will you have to complete the project? If the client needs something in two days then ray traced animations are probably out of the question. If you are doing things for yourself then you can take all the time in the world. Of course, a program about the newest software won't be very valuable three years from now while a documentary about the 1989 San Francisco earthquake might be interesting forever.

How much money can you spend on the project? If it is your own money then this is one of the most important questions you can ask. If someone else is footing the bill then you will have a budget of some sort. If you only have a few hundred dollars then location shots in the Figi Islands are probably out of the question.

How should you spend the money that you do have? Depending on the project you might want to spend most of your money on travel, props, talent ('talent' in this case refers to the actors), setting up a studio, special hardware or software, etc.

Are you comfortable with the project? This might seem like a strange question to ask but if you aren't comfortable with a project then at best it won't be very enjoyable and at worst you might ruin everything. Professionals are used to working on projects of someone else's design but if you are squeamish do you really want to do a program on autopsies? If you are puritanical do you want to work on a pornographic video? If you are honest do you really want to work on propaganda pieces?

How are you going to approach the project? Before you get down to detailed planning you should decide how the finished project will look and what style you are going to use. Is it going to be straight forward, artistic, flashy or simple. Will there be a lot of location shooting, studio work, interviews, narration? Will it be a step-by-step format, montage or unfolding story? Will you be needing actors, animation, special effects? All of these elements should be considered before you get started.

Given the time and resources can you complete the project? Most of us desktop video producers would love to work on a grand-scale project. Most of us also have grand ideas about things we would like to do. However, the truth is that most of us don't have the resources to do most of these projects. The biggest limitation is likely to be the budget (with enough money you could do just about anything). There can be other factors too. Time, getting actors, getting permission, bad weather, legal problems, and on and on. There could be a thousand reasons why you couldn't complete a project. If you are determined and resourceful enough you may be able to overcome many problems but there are some things you just can't do anything about. If you are thinking about a project for someone else and you know that you just can't do what they want it is probably better to let them know in the beginning. That way you can either re-think the project or try something else. Even if you lose the project people won't be angry with you.

## Outlining

There are many other things you should consider before embarking on a project and you will find them yourself. Let's assume that you have asked all the right questions and came up with all the right answers. You have the time, and the money you need (actually, you will never have all the money you need), you know your audience and how they will be watching, you know how long the production will be and you are comfortable with the idea. The next step is to do a quick outline of the project. An outline can be as simple or as detailed as you like but the more time you spend on an outline the less work you will have to do later. An outline will help you organize your thoughts and force you to look at 'the big picture.' You may have a hundred images and ideas running around in your head but if they aren't connected your production will suffer. List the major points that you want to get across and the order they should be presented. Then go back and fill in the secondary points. Then go back and jot down the best ways to present those points. If you want to you can even go back and make a quick list of the scenes you will want. These don't have to be very detailed right now. For example, if you were doing a video on baking a cake your first outline might be:

```
Talk about cake
assemble ingredients
mix ingredients
cook cake
frosting the cake
serving the cake
```

The second level outline might be:

```
Talk about cake
    Describe cake
    Where recipe came from
Assemble ingredients
    List ingredients
    List substitutions
etc.
```

The next level outline might look like this:

```
Talk about cake
    Describe cake (voice over)
    Where recipe came from (voice over)
Assemble ingredients
    List ingredients (text screen)
    List substitutions (text screen)
```

The final outline might look like this:

```
Talk about cake
    Describe cake (voice over) - finished cake shots
    Where recipe came from (voice over) - grandmother shots
Assemble ingredients
    List ingredients (text screen)- overlay text on ingredient shots
    List substitutions (text screen)- transitions with new text
```

If you were doing a murder mystery some of your final outline might look like this:

```
The Murder
    victim tries to escape (night) - shots of victim running
    victim shot (dead end alley) - shots of alley, gun firing, victim
                                                                dying
    murderer leaves clue (murderer running away) - dropping gun and
                                                               flower
Detective brought in on case
    gets phone call (detective's office) - phone, detective, writing on
                                                                  pad
    detective finds clues (alley again) - walking around, finding
clues,
        etc.
```

If you go this far then you will begin to see what kinds of things you are going to need. In the cake example you will need some photos of granny, a titling program and perhaps a presentation program for the

197

transitions, some real ingredients and a finished cake. In the mystery example you will need some actors, some props (gun, flower, phone, pad and pencil) some sound effects (gun firing) and obviously some location shooting which might require extra lights or special filters to make it appear to be night.

Another thing that an outline will do is give you some idea of the order you will have to do things when shooting and editing. In the cake example you'll have to take some extra shots of the finished cake before you show how to serve it. Otherwise you won't have anything to edit into the beginning later on. In the mystery example you might want to bring everybody to the alley location to shoot both scenes at the same time. You might even want to shoot the detective parts of the alley scene in the afternoon and then the murder scene afterwards, when it starts to get dark.

## Scripting

Once you have your rough outline done it is time for the detail work. Not too many people enjoy detailed scripting but it is well worth the effort. A detailed script should include all the shots in every scene, all the locations, all the camera angles, all the camera movements, all the actor movements, all the words that are going to be spoken (by actors and voice overs), all the sound effects, all the music, all the transitions between shots and basically every detail that you can think of.

There are many ways that you can put together a script and you may even end up with more than one script. Some productions may call for a story board treatment where there is a sketch of every camera shot (from the camera's point of view). Some productions may have a very sparse script (particularly an interview type production which may only have a few camera shots and no pre-prepared dialogue. Some scripting will be mostly narration with a few notes on the actual shots to be used while others may have detailed shot descriptions and very little dialogue.

It might be easier to keep a loose leaf notebook devoting a separate page for each scene (more if necessary). At a minimum your script should include the following for each scene:

1. Scene number, perhaps a scene title, a brief description of the scene that includes location, setting (special lighting) and list of characters.

2. A list of the visual information in one column (including transitions, camera angles, camera movements, object movements, actor movements, animation, titles, special effects, etc.).

3. A list of the corresponding audio information in the next column (including dialogue, voice overs, music and sound effects).

4. A column for the duration of each shot.

5. A space for notes that might include props needed, number of takes, problems encountered, extra shots not in the original script, things to be corrected in the studio (unwanted sounds for example) or things to be done later.

An example Scene page might look like this:

```
Beast at the foot of the stairs                                  page 18
Scene # 12
Scene Name: Fred gets eaten        Setting: Basement - Dark and spooky
Need: Fred, monster, monster hand, spider web spray
Shot#      Video                            Audio                  Time
  32  Fade in from black                                           00:20
      Long shot of Fred at top of stairs   scarey music starts
      Slow zoom in as Fred looks around
      then reaches for light switch
  33  Cut to Tight shot of hand trying     FX-switch clicking      00:05
      switch a few times
  34  Cut to 3/4 shot Fred looking angry   Fred: "Damn lights!"    00:05
  35  Cut to medium shot of darkness                               00:05
      under stairs. Slow zoom in
  36  Cut to long shot from foot of stairs bring up music a bit    00:10
      Fred starts down stairs
  37  Cut back to slow zoom of darkness                            00:05
      under stairs. FX add glowing eyes
      with paint program later
  38  Cut to tight side shot of shoes      FX-creaking steps       00:10
      going down steps. Pan and tilt to
      follow
  39  Cut back to zoom under stairs        FX-monster growl        00:05
      FX of glowing eyes
  40  Cut to tight shot of shoes                                   00:05
      stopping in mid-step
  41  Cut to tight shot of Freds face.     Fred: (voice cracking)  00:15
      He looks nervous and scared. Licks   "Is anyone down here?"
      lips then speaks.
  42  Cut to dark under stairs.            Music gets louder       00:02
      FX of glowing eyes
  43  Quick cut to monster lit                                     00:02
      momentarily by lightning
  44  Quick cut back to dark under         FX-thunder              00:05
      stairs. FX of glowing eyes
  45  Cut to tight shot of Freds face.     FX-thunder roll         00:10
      He jumps at sound of thunder.
  46  Cut to shot of shoes from behind.    FX-monster breathing    00:20
      Shoot from under stairs. Slow zoom
      in on shoes. FX of monster-eye-view  Fred: "Hello?...Hello?"
  47  Cut to side shot of monster hand                             00:04
      reaching.
  48  Cut to long shot of Fred paused on   FX-monster growl        00:05
      steps.
  49  Cut to tight shot of monster face.                           00:03
  50  Cut to tight shot of Freds face.     Fred: "Oh shi...."      00:03
  50  Cut to monster hand grabbing ankle   FX-monster roar         00:02
  51  Cut to full shot of Fred from bottom Fred: "IYYEEEEEEE!"     00:05
      of stairs. He falls forward toward
      camera.
  52  Cut to black                         Music ends abruptly

Notes: Shots 35,37, and 39 are all one shot cut apart later. Shots 43, 44, and
45 are all one shot but flip lights on for an instant (for lightning effect on
shot 44.)
```

This might seem like a lot of work but most of the time it will be worth it. When you finish the detailed script you might want to go over it one more time and do a shooting order list. Like in the mystery outline example above a shooting order list would have all the scenes to be shot in the alley, all the scenes to be shot in the detective's office, etc. This will help organize things and save you time and money in the long run. This list doesn't have to be more than a page or two.

While a script for a feature movie might be hundreds or thousands of pages long your scripts might be only a page or two. And of course if you are doing a production for yourself you don't have to follow a script to the letter. But if you are doing a production for someone else and they have already approved a script then they might not appreciate your taking a creative license without letting them know.

An outline, a detailed script and a shooting order list will help you determine how long things will take and writing it all down will give you an idea about the kinds of actors, props, music, sound effects and special hardware or software you might need. They can also point out potential problems you might encounter. And it is much easier to make changes to a script then to go out and shoot new material or try to patch things up in the editing studio.

## Production

Now that you have all the boring stuff done you can grab your camcorder, go out and start shooting right? Not quite. First you should line up your actors and assistants, props and equipment. Next you might want to scout your locations to check for camera angles and to be familiar with the area so that when everyone arrives you won't be wasting time looking for outlets or deciding where to put the camera. Jot down a list of the special things you will need during the shoot. Not just lights, microphones, tripod, etc., but things like a step ladder for climbing that tree, a sheet of plywood to put the tripod on, rope to block off an area, beer cooler to sit on, etc. You might also want to check to see if you need special permission to shoot there. Even if you are going to be shooting in your own home you might want to check camera angles and get permission from your spouse.

When the day of the shoot arrives go over your check list two or three times before heading out. You do have a check list don't you? You will develop your own checklist but it should include even the obvious items. Here is a short checklist you can work from:

Camcorder, plenty of tape plus an extra cassette or two, lenses, filters, microphones, cables for microphones, spare batteries for microphones (if they need them), batteries for the camcorder and any other equipment that needs them (they are all charged right?), extra batteries for everything, the battery charger, tripod, lights (list all of them), tripods for the lights, extension cords (for the lights), three prong adapters (for the extension cords), duct tape (great for quickly holding light things in place, pens and paper (for notes), your script, the props (check the script one last time), first aid kit, tool kit, snake bite kit, cork screw, money, wallet, (you did get the car filled with gas didn't you?), carrying cases for everything, copies of your permission (if you have written permission that is), and anything else that you wouldn't want to drive back to the house for.

Before you leave, you might also give your actors and assistants a call to make sure they don't oversleep. Of course, if you are shooting by yourself in the backyard then you don't have to do any of this (except check your batteries).

Once you are on the set or location the actual shooting is up to you. There is little that I can tell you about how to shoot in different situations. About the only advice I can give is to get some books on production techniques. You should already be familiar with all your equipment (how to set it up and use it). One tip is to practice your shots before you tape them and shoot more than you think you will need. It is also a good idea to leave extra seconds at the beginning and ending of each shot to give you room for editing. In the end it just takes practice, skill and planning.

Once you have shot your scenes label your tapes and head back home. If you did everything right then you should be ready to start bringing the computer into the picture.

## Post-production

Back in the studio you can get to the real work of putting it all together and this is where the computer comes into play. Technically, anything that isn't done with a camera is either pre-production or post-production. There are elements of computer generated video that probably should be considered studio work and not post-production (like animations for example) but it doesn't really matter in a desktop video environment.

Where it does matter is in the order you do things. Because most of the computer generated or computer enhanced scenes will be separate scenes all by themselves, you should do most of them before you get down to the actual editing. And unless you are doing music videos you will want to save the audio sections (sound effects, music and voice overs) for the very last step. That way you will know exactly when and how long the audio will come in.

The only exception to all of this is when you are mixing computer effects and live action (like Roger Rabbit). In that case you will have to plan your production scenes carefully to coincide with the animations.

If you did your detailed script then you already know what shots will be completely camera generated, which ones will be completely computer generated and which ones will be a combination of both. You should have the same kind of shooting order list for your computer sections as you would for your camcorder sections. This way you can do all of your titling tasks at the same time, all your digitizing or frame grabbing, all your animations, all your sound effects and music, etc. There are two reasons for doing this. First, it saves time when you work on the same kinds of computer elements. You can also re-acquaint yourself with the software and hardware for a specific task all at once. Second, many of the hardware and software products are not compatible with each other so you may have to exit the program and re-wire things in order to go to another type of activity (genlocks and digitizers usually won't work together because they both use the same ports on the computer).

There really isn't any set order in which to do things with the computer but some things may build on others. For example, if you are going to use digitized or grabbed images for backgrounds, for titles or animations then you should do the digitizing first. In general, you might want to do your computer and video tape editing tasks in this order:

1. digitizing or frame grabbing

2. paint program, rendering and ray traced graphics

3. animations and page flipping

4. titling tasks

5. image processing (if necessary)

6. presentation scripting

7. computer to tape transfers with genlock

8. computer overlays with genlock

9. video tape editing

10. music and sound effects composition

11. music and sound effects transfer to audio tape

12. voice overs transfer and mixing to audio tape

13. audio mixing and dubbing to video tape

Let's take a look at each of these elements and how they fit in to your production.

Digitizing and frame grabbing - You might want to do all your digitizing and frame grabbing first because there are many other computer applications that include digitized images. Just be sure that you are using the interlaced, over scan modes or you will have to use an image processing program later on. Even if you are thinking about DVE type effects with a digitized or grabbed image it is probably better to use an image processor later on a full sized image. Digitizers and frame grabbers also don't usually work with other hardware (like genlocks) at the same time so you will have to do your digitizing first and then your genlocking later.

Paint Program, rendering and ray traced Graphics - If your production requires a static graphic created on the computer then you should simply create the graphic first, making sure that you are using interlaced and overscan modes and not using illegal colors. If the graphic is going to be overlayed on video be sure you know which is color zero on your palette. Don't use that color in your image and don't forget to 'paint' the parts where you want the video to come through with color zero. Save the graphic on disk right away and later when you have your genlock set up to do all your transfers at once you can put it on video then. You could try to fit it in during the editing process but that can be a little tricky. It is much easier if you tape the shot first and edit it in later.

If you are going to be using color cycling effects then obviously you will want to be checking your work as you are creating it. When it looks the way you want then save it. There will be times, however, when the only way to check an effect is to put it together and look at it then. This is going to be mainly trial and error and it is a good idea to keep notes of what worked and what didn't.

If you are going to be modifying digitized or grabbed images or even titles, then you will have to do them first and then bring them into the paint program. If you are going to be using a paint program to generate custom pointers be sure to save them as brushes and keep in mind that you will probably have to use them with the original paint program when transferring to tape.

Rendering and ray tracing is going to take a lot of time so keep that in mind from the beginning. You could get some of your other production work or video tape editing done while you wait for images. It is a very good idea to be familiar with the software before you embark on a big project that requires rendered graphics or rendered animations.

**Animations**

The order in which you do your animations depend on the type of animation and what other elements might be included. If your animations are going to be very long (anything over 60 seconds is long for an animation) or involved (anything with more than one shot is involved for an animation) then it is a good idea to do a script for them scene for scene and shot for shot. While it is possible to do single frame animations directly to tape it is much easier to create ANIM files and transfer entire scenes to tape later. ANIM files can be manipulated and modified much easier than altering an animation already on videotape.

If you are doing claymation or silhouette animations then you might have to digitize or grab a number of your images and save each one to disk. Then go back and put them in an ANIM file with a page flipping program and check your results. Then go back and grab some more images and save them. Once you get a feel for how much to move things you will be able to grab a larger number of images in one session before putting them into an ANIM file. You probably won't be able to grab ten minutes worth of images unless you have a lot of diskettes or a very large hard disk drive. ANIM files take up much less space than a string of images. You might also consider breaking the animation up into separate ANIM files and then editing them together after they have been transferred to video tape because the length of the ANIM files you can play back at one time depends on the amount of memory you have in your Amiga. Some frame grabbers will create ANIM files for you as you go along which can save you a lot of time and disk swapping. Just be careful that you don't end up creating an ANIM file that is too big to play back on your machine.

If you are using an all-in-one animation program then you can go ahead and create the whole thing before transferring it all to tape. The programs should let you know when you are getting close to your memory limit while you are creating the animation. The same holds true for paint programs that have animation capabilities.

If you are doing paint, rendering or ray traced animations then you might want to storyboard or pencil test them with an all-in-one animation package first, just to see if the actions are right. If it looks good then you can start. How you animate things with paint or rendering programs depends on the packages you are using. If they can produce ANIM files automatically then you will save a lot of time. If they don't then you will have to use the same technique as with claymation animations (do a series of images, put them in an ANIM file with a page flipping program, do another series, etc.). Where these animations get complicated is in the fine tuning and touch up work. Some people create their objects with Sculpt because it is easier to design with. Then using Interchange they transfer their images to Silver because they like the way Silver renders. Then they clean up the images with Deluxe Paint. Then they create an ANIM file with Page Flipper Plus. Then they might adjust the ANIM file with Animation:Editor. Then link the ANIM files together with TV*SHOW, Deluxe Video, The Director or Lights! Camera! Action! Then they transfer it to video tape. You get the idea. It is complicated, time consuming and frustrating. Many times you will run into compatibility problems, resolution and mode problems and quirks with the different software packages. Don't expect things to work smoothly on your first attempt (or your second or third or fifth).

**Titling**

In this case titling covers anything that has to do with text on the screen (although a logo or static graphic might fit into this category too). You should treat a titling sequence just like you would any other scene. Create a small script just for the titling scene, listing each shot (or screen), transitions, timing, etc. If there is only one text screen then you don't have to do this but there should be a scene page in your main script for it so you don't forget it.

The next step, obviously, is to decide what words you want to appear on the screen in each shot. Next you want to decide what will be in the background. Will you be using backgrounds generated by the titling program, digitized images, grabbed images, paint program images or will you superimpose the text onto video? You should have already given this some thought during the scripting process so that if you need them you will have shot extra video just for the titles. Another consideration is if you want other elements on the screen at the same time like animations or reduced 'snap shots' from the production. If you are going to use digitized, grabbed or paint program images as backgrounds or extra elements then you should do those first. If you are going to add animations, however, it may be easier to bring static text into an animation as a background than to try to bring animations into a text screen.

If the background is an IFF image (including digitized or grabbed images or images made with a paint program) then most titling packages will let you import that image as a background without any trouble.

If you are using a screen/text/display type titling package you can go ahead and create each one of your screens, decide on the backgrounds and transitions. Preview your creation and then save the file for later transfer to tape. If, on the other hand, you are using a screen/text titling program you should create your individual screens first and save them to disk, then move on to your presentation program to tie them all together. In both instances you should double and triple check your spelling during the screen creation process.

If you plan to superimpose your titles on video then during the screen creation process you should make sure that you are using color zero for your background. You might also wish to make a trial run or two with the video source genlocked in just to be sure your font colors will work with the video background. You don't want to have 20 pages of white text on top of a white sky. This is why outlining and shadowing is important.

## Image Processing

Image processing is usually done when you have all the parts finished and before you start putting them together. There are no hard and fast rules that say when you might want or need to do something with your graphics. Sometimes it is best to work on digitized or grabbed images right after you finish them while you still have an idea what you want to do. Sometimes you might find that you have to work on images during or after you do an ANIM file or presentation script. Sometimes you will know that you have to manipulate images before going on to another process and sometimes you find your problems when you play back a sequence for the first time. Usually you find a need for image processing programs when you start mixing graphics and realize that the palettes don't match from one to the next.

## Presentation Scripting

When you have all your pieces ready you can start the presentation scripting process. The reason I call it scripting at this point is we aren't ready to transfer it to tape yet. You might consider this stage a rehearsal. If your presentation is going to involve a lot of transitions, animations and other graphics or sounds then you should write down a quick script that contains the image file names, the transitions you want and a rough idea of the timing. You will probably want to move your files around so that they are all on one disk or just a few disks. This is where a hard disk drive comes in very handy. At this point you might want to jump out of sequence a little bit. If you are planning to have computer generated audio in your presentation then you should go ahead and do that before getting to the presentation phase. If you are going to add the audio later then you don't have to worry about it now.

When you have all your files in one place then you can start putting together the presentation. Try to stick to your written script at first. Later, when you play it back for testing you can go in and adjust things if they don't look right. When all the timing and transitions look right you should do a final run through or dress rehearsal. This is where you might find that the program hangs up waiting for you to switch disks or palettes shift during transitions from one graphic to the next. You can go back and fix your images with an image processor. If you can't overcome the disk swapping problems by moving files around then you might have to break up your presentation into parts and edit them together later. When everything looks and plays back smoothly with no surprises, save the script, make a note of the disks you need in which drives and go on to the next task.

**Computer to Tape Transfers**

You should have a list or lists of all the things that you want to move onto video tape before you begin. You should group the things into categories based on which software you need to display or play them back on the computer. That way you can do all the paint program images at one time then the titling sequences, ANIM files, presentations, etc., and not have to jump back and forth between programs. When you have all the parts ready, turn everything off.

**WARNING!!!**

Trying to hook things up while the power is on is probably not dangerous for you but it is a good way to destroy your computer, genlock, VCR and camcorder all in one shot!

When everything is turned off you can hook up your genlocking unit. Connect the genlocking unit to the computer following the directions to the letter. Also connect the computer monitor with the RGB connection on the genlock. This way you can see what the computer is doing before it gets to the VCR. Even if you are not going to be superimposing images on video most genlocking devices work better with an external sync source so feed your camcorder's 'video out' into the 'video in' port on the genlock. Be sure to leave the lens cap on the camcorder to avoid burn-in or unwanted images sneaking in. Now connect the genlock's 'video out' to the VCR's 'video in' port. You will also want to feed the video signal from the VCR into a monitor to see what the VCR is 'seeing'. When everything is connected and you have checked everything twice then you can turn everything on. The order in which you turn things on probably isn't that important but occasionally it can make a difference so if you seem to be running into problems try a different order.

Put a fresh tape in your master VCR and reset the tape counter, load up the software, load up the right files and check to make sure you have the correct disk or disks in place. Set your genlocking unit to 100% computer 0% external video, start the VCR and give it a few seconds to get up to speed before starting your stop watch. Be sure to tape more than you will need to give you some editing space later on. It is a good idea at this point to do a test run on something just to make sure that everything is hooked up properly. If it looks good then you can go ahead and transfer the rest of the images in that category.

When you are finished with a category of transfers you should also go back and check the tape right away, before turning off the computer or going to the next transferring job. This is just to make sure that the transfer went all right. If it still looks good you can go on to the next category of transfers. If it doesn't look right then you may have to re-tape or even go all the way back and change things with an image processing program. With any transfer you should give the VCR a few seconds to get up to speed. When you have successfully transferred an image you should jot down the tape counter numbers to make it easier to find later on.

**Computer Overlays**

Follow the same procedure as above but this time instead of the camcorder you should connect your slave VCR's 'video out' to the 'video in' port on the genlock. That is unless you plan to overlay your graphics on top of live video from the camcorder or you are using your camcorder as the slave unit in which case leave the camcorder connected to the genlock. Go through the same steps you would for a straight transfer except that before you start your master VCR you should find the video on the slave VCR or set up your live shot that you plan to overlay on top of. You should then check your genlock settings and make sure the overlayed graphics work well with the video. If there are problems here (like color bleeding or difficult to read text) you should go back and correct them before trying to transfer to the master VCR. If the overlayed graphics and video look right you can then determine what your genlock settings will be, including the in and out transitions. Make a note of these settings. You should then time the elements of your scene with a stop watch. Find out how long the video section is and how long your computer images will run. You should also write down all your cue points if transitions depend on changes in the video. Once you have done all of that it is a good idea to do a complete dry run. Rewind and cue up the slave VCR, cue up your computer images and give it a try. If everything went smoothly you can go ahead and transfer everything to the master VCR.

All this may seem complicated (and it is) but with practice you should get the hang of it. One thing that you will find is that in the middle of one of these transfers you don't have time to read your stop watch. Most video tape editing people reach a point where they can count off seconds in their head fairly accurately. If it seems too complicated you might re-think the overlay idea and try putting your graphics on digitized or grabbed images instead.

If you are worried about extra generations you can complicate this whole procedure even further by trying to edit in this mess onto your master tape while you are doing it live. If you want to try it you should have four or five hands, the ability to keep track of six or seven different things simultaneously, a lot of courage and a lot of luck.

## Video Tape Editing

Video tape editing is as much art as it is science and the only way to get good at it is to practice. You should go back and re-read Chapter 2 on the basic types of editing (assemble edits and insert edits). The first thing you do before launching into editing is to put together an EDL (edit decision list). This is a list of all the edits you are going to make, the in points, the out points, times and where all the parts are located on which tapes. The best idea is to find your edit points, time everything with the stop watch, go through a dry run and then give it a try. With a good script, an EDL and a lot of work you should be able to put together the shots you want in the order that you want them. After trying to edit a handful of scenes manually, you will discover why people buy editor/controllers.

## Audio

Once you have your production edited together you can go back and time scenes that require extra audio. When you know how much time you have to fill you can start composing your music and putting together your sound effects. The music side of things can be as complicated as you want to make it or you can use one of the idiot-proof programs. Apart from matching the moods of the visual you also have to match the timing. When you have matched both the mood and timing you might want to transfer the computer generated piece to audio tape because it can be easier to transfer from audio tape to video tape than from the computer directly to video tape.

The transfer from computer to audio tape can be done straight from the Amiga stereo audio outputs into a tape deck and control the volume on the tape deck. But a better configuration might be from the Amiga into an amplifier first and then into the mixer then into the tape deck. You may want to leave your Amiga audio output connected into your sound system amplifier auxiliary inputs and leave them there to give you stereo sound all the time anyway. You can connect your 'tape out' ports from the amplifier to the mixer. You would connect your tape deck to the tape in and tape out connections on your mixer. This will give you more control over the volume. If you have enough connections on your mixer you can run the audio out from your slave VCR and your stand microphone into the mixer too. And finally, run your main mixer output to your master VCR 'audio in' ports.

Recording audio from the computer is pretty straight forward. The only problem you might encounter is hooking things up to the ceramic inputs rather than the magnetic inputs on the mixer or vice-versa. If you get a terrible buzz when you increase the volume on the mixer try switching from one input to the other. You will probably also want to run another main output from the mixer back into the amplifier so that you can monitor your recordings. You have probably realized by now that you don't have enough inputs and outputs on your mixer. You don't have to go out and buy another mixer just plan your activities so that you aren't trying to do five things at once and label your cables so that you can switch things around easily. The trick is getting all your audio on tape before you get to the audio dubbing to video.

A second tape deck can come in handy, particularly when you want music and voice overs and/or sound effects at the same time. One trick that you might use is the fact that a VCR will record sound too. So, if you think about it you already have three audio recorders (two VCRs and a tape deck). You could record the computer music onto one VCR then feed the VCR and your stand mike into the mixer while you are recording your voice overs onto the tape deck. Once you have all your extra audio onto tape you are ready to do your audio dubbing.

The simplest method for audio dubbing is to time your edited production very carefully, find all your 'start tape deck' cue points, your 'tape in' points and your 'tape out' points. Run both your slave VCR audio and your tape deck into the mixer and the main audio output from the mixer into the master VCR. Start recording on the master VCR, start the slave VCR. As your 'start tape deck' cue points come up start your tape deck, as your 'tape in' cue points come along bring up the audio from the tape with the mixer, then as the 'tape out' cue comes along bring down your tape deck audio with the mixer. If this doesn't sound simple that is because it isn't. One way to make it a little easier is to break it up into parts so you only have to worry about one piece of extra audio at a time. At the end of a scene everything is quiet. This is the audio equivalent of a fade to black. If you leave yourself a little room for error and don't try to match the extra audio with any specific video images things will be much easier. Another method might be to do multi-track audio mixing and then sync everything at once. This is the method most films use but it requires a lot of expensive equipment. Finally, you can completely replace all the audio. This might be the easiest for educational or industrial videos where you don't have to show people talking.

If you have a stereo VCR with audio dubbing capabilities then you can leave the original audio on one track and put your extra audio on the other track. If you have a mono VCR with audio dubbing then it might be worth trying to sync your audio. First, do your video editing. Second, make a dub of the final tape (don't worry about the video quality). Third, find starting cue points so you can get the dub copy running in step with the master copy. Fourth, do an audio dub on the master feeding the slave audio (the dub) and the tape deck into the mixer with the mixer output going into the master VCR. It might take a few attempts to get everything in sync but without special equipment audio matching is going to be mainly hit and miss.

If your master VCR doesn't have audio dubbing then you really only have two choices. Go to another generation to add your extra audio or try adding the extra audio during the editing process (which can be a nightmare).

Audio dubbing can be just as tricky (and frustrating) as video editing but it may be worth it in the end.

## Summary

I know that I haven't covered everything in this chapter and I'm sure that you will find other ways to do things. Remember that most of the ideas here are just suggestions and outlines that you can follow or not. You may also run into problems that I haven't mentioned. It is fortunate that there are so many companies making different products of use to the desktop video producer but that also means that some things won't work with others. It is always best to try something out before you sell the idea to someone else. It is also a good idea to start out with just a few simple things rather than trying to do everything in your first production.

# Chapter 12

## Advanced Techniques

# Chapter 12
# Advanced Techniques

While most of this book has been devoted to getting started with desktop video, I have mentioned a few of the more advanced techniques along the way. This chapter is again a kind of introduction into more sophisticated applications. Most of the time you could think of these as extensions of the ideas that have already been covered. Advanced titling, advanced graphics and animation, advanced genlocking and digitizing. But there are other less obvious uses for desktop video and ways to get professional results with your desktop video setup.

In a way you can divide professional desktop video applications into two categories. Those that use the computer in a video application and those that use video in a computer application. Of course there may be instances where you have a little bit of both. If, for example, you are putting together a business presentation you might use video as a way to gather information for a computer demonstration that is later transferred to tape.

When the Amiga was first thought of as having uses in video a number of companies bought them, tried them and decided that they just couldn't get broadcast quality from them. In the beginning that was true. But it wasn't just the Amiga. The biggest problems were the genlocking devices available. They produced signals that were barely capable of putting an image on a screen and were far from NTSC standards. Now there are much better genlocks for the Amiga that can produce nice clean signals. The next problem was (and still is) the fact that the Amiga can generate illegal colors. Most professional video people never considered that a piece of equipment might be capable of generating illegal signals. When they saw white levels climbing off their scales (or the poor picture quality that results from oversaturation) they assumed the Amiga was at fault. They were right to a degree. But they probably didn't try different color combinations either.

Another problem in the early days was the software for the professional user. It either didn't exist or it wasn't very good. Now the variety and quality of the software has improved a great deal and there are professional level packages out there that perform very well. Finally, the users just didn't understand computers and didn't want to. Unfortunately, there isn't much you can do about that. If you don't want to spend the time to learn how to use the computer yourself then you can either hire someone who is willing or don't expect professional results.

These days you can buy professional quality genlocks, if you have read this book you are aware of and can avoid illegal colors, the software is much better than it used to be and if you spend a little time learning to use the computer then there is no reason why you can't use the Amiga in broadcast situations. Hopefully, I have raised your curiosity level enough so that you will want to learn the computer to do some of the things I have talked about. I won't tell you that the Amiga is perfect in every video situation so don't expect miracles. But I think that the computer/video combination can be a valuable tool even in broadcast environments. If nothing else, desktop video can be an inexpensive way to work out your ideas before you go to other, more expensive equipment to get your professional results.

I can't cover all the advanced desktop video applications in one chapter so I will just mention a few things. Things like animation, downloading satellite information for weather maps, MIDI music composition, interactive video and laser disks are all advanced areas where people have spent a great deal of time and money developing tricks and techniques. There are archaeologists who use digitizers to accurately record and map progress on digs. There are astronomical illustrators who work with NASA creating computer paintings and animations based on scientific data. There are medical research firms using Amigas and infrared video cameras to look at body heat patterns. All of these applications fall into the advanced users category but all I can do is mention them in passing. Here are a few of the more common applications you might run into and some ideas for projects of your own.

# Titling

Perhaps the most obvious advanced application for desktop video is in titling. There are a number of cable companies and broadcast people who use the Amiga as an inexpensive yet very flexible character generator. Since you have almost unlimited fonts without having to change your equipment the Amiga is a very attractive alternative to buying a broadcast character generator or for use as a secondary character generator. Another use for the computer in titling is in creating custom title screens. In both of these instances the thing you have to watch out for is using illegal colors. If you are thinking about doing titles for broadcast or cablecast use be sure to check all your screens with a waveform monitor. You will also want to use anti-aliased fonts with outlines to make sure you get the sharpest letters that are easy to read. You also want to check all of your transitions to make sure that your palettes don't shift and the timing is right. Of course you will want to use the best genlock you can get and test everything before you go on the air.

Fortunately, computers, like most solid-state electronic equipment can be left on for weeks at a time without wearing out. But if you plan to leave the computer on and unattended for long periods you should turn it on and leave it on for an hour or two when you first install it to make sure there aren't any faulty chips. You must also make sure that the computer is well ventilated. If the computer begins to act strangely (most likely it will just lock up and have to be re-booted) then check for chips that feel hot. You might also check the genlock. If the problem occurs only rarely it may be your power source. While you might never notice it most power sources fluctuate quite a bit during the day. In some areas this fluctuation is just enough to effect computers. In this case you may need a line conditioner (a device that levels out a power source).

In a broadcast situation you might not want to be relying on computer generated titles in a live situation so transfer everything to a professional VCR first. That way you can check your quality ahead of time.

# Educational Uses

There are thousands of ways to use a desktop video setup in education. The two most common applications will probably be putting together educational videos or letting students put together their own videos. You can even kill two, three or a whole flock of birds at one time if you have students making educational videos for other students. Not only will the students be excited about the projects (which obviously inspires learning) but they will be learning a number of different skills. Planning, organization, video, computers, interaction, writing, graphics and half a dozen other disciplines. On the other hand, if you are doing your own educational videos then apart from the usual steps in producing any video there are other considerations. You should try and make things entertaining as well as informative because there will be many occasions when your audience is being forced to watch. This is where video can be a great help. Reticent students will be more inclined to watch a video than listen to a lecture or read a book. Special effects, graphics and animations will all help to bring the viewers in. The best advice I can offer is to give as many real-life examples as you can, watch as much educational television as you can and emulate the programs that work. Obviously, if you find something boring then most students will too.

## Business Presentations

Computers are great for manipulating information in many ways. Most of the time in a business environment it is the manipulation of numbers or text. One of the problems ,however, is how do you share that information with other people? The computer can generate the information but getting the information out of the computer can be a problem.

The easiest and most traditional way to share that information with other people is to print everything on paper. If you wanted to get fancy you might even use a desktop publishing package to make your printout look more professional (there are a number of desktop publishing packages available for the Amiga if you are interested).

Paper does have it's obvious advantages but there are disadvantages too. Paper is a static medium so you can't really show the evolution of your information (why you did things one way rather than another). It is also difficult to try 'what-if' exercises with a list of printed figures. Paper is also a solitary method of communication that depends on the reading speed of the people you are trying to communicate with. This makes it difficult to share information with a great number of people unless you have a lot of time. Paper can also get to be expensive (calculate the expense of giving a twenty page report to an audience of 100 people).

There are two methods where you can use the power of desktop video in a business presentation situation. The first method is doing your presentation on the computer itself. The second method is to transfer your information to video tape. Both of these have advantages and disadvantages.

In either case you should go through all the steps outlined in Chapter 11. Plan your production, outline how it will appear, script as much as you can, gather all the elements, then put everything together and test it. In the case of a live presentation you will also want to add three more steps. Practice, setup and the presentation itself.

When planning you have to decide who is going to see the presentation (chairman of the board, prospective customers, new employees, etc.).

where they will be seeing it (conference room, company picnic, your office, lunch room, etc.), what information you need to present (financial secrets, products, people, plans, etc.), what form the information will take (lists of numbers, words, pictures, interviews, etc.), how long you want it to run, will it be all computer generated or a combination of computer and video, and finally, you should decide if another medium might be more effective. Your outline should cover the order in which you want to present the information. This will help make your presentation more logical and coherent. Your scripting will help you determine what things you will need and how long you might expect everything to take. Gathering the elements might mean taking photographs and then digitizing them, grabbing frames from a company video, gathering numbers, figures, charts, etc. Putting everything together means doing all the graphics work, creating screens, and finally, putting everything into a presentation package and trying it out. You will also want to practice the entire presentation to see how it will go and where there might be problems. Try to anticipate questions and interruptions. Setup means making sure that all of the equipment you need is going to be in the right place when you need it and that it will all work together.

If you are going to be doing the presentation on the computer then you will either have to bring everyone into your office to huddle around the computer or bring your computer to the person or persons and give your presentation there. In either case you will have to overcome the display problems. The simplest way to do this with a small group is to elevate the monitor as much as you can. The next way you can do this is by using a larger monitor or video projector. If you want to connect your computer to a large TV then you will need an encoder (or use the encoder built into your genlocking device). If at all possible you should try out the equipment well before your presentation because when it comes to doing anything in front of a group that involves electronic equipment you can almost always count on something going wrong. If you can't try it out first then make sure that you give a very detailed list of the things you will need, triple check everything ahead of time, bring extra cables, power cords, adapters, backup copies of all your disks and show up extra early to set things up. It is also a good idea to work out a contingency plan in case the worst happens. Bring your script with you, it may end up becoming your notes if you have to talk.

If on the other hand you are going to be putting the entire presentation on video tape then there are other considerations beyond the ones above. While you may have more control over the production itself you may have less control over where it will be shown and who might see it. Therefore it is probably not a good idea to put sensitive information in a video. Since you may not be there to talk through things you might want to put more detail and explanation in the video.

## Demonstrations

Similar to presentations a demonstration should be planned, outlined, scripted and the rest. The biggest differences between a presentation and a demonstration is your audience, the information and the setting. While a demonstration may include a lot of similar information it should be presented with the audience and setting in mind. Most people won't want to sit through a lot of numbers and figures unless they have a reason to. If your demonstration is a store display or any area where there is a lot of traffic (like a trade show) you will need to make your demo eye catching and try to get your message across quickly. You also have to consider whether or not you are going to be there to control things or will this be a self running demo. If it is a self running demo then decide if you want it to be interactive. (Interactive demos can be very effective but also harder to put together). All of these elements will determine how you put the demo together. If it is to be a self running, non-interactive demo then video tape can be an ideal solution. If, on the other hand, it is to be a self running, interactive demo then you will have to use the computer.

## Live Performances

You could think of live performance as a form of demonstration (with a small stretch of the imagination). A demonstration of your performance skills. There are a few companies that produce products designed to incorporate video, graphics and computer control over MIDI devices, lighting and almost anything. Right now the most flexible and one of the most intriguing software package designed for live multi-media performances is Elan Performer from Elan Design. It is a kind of presentation package that lets you load and shuffle animations, graphics and even music files. Used in conjunction with the Live! real time digitizer from A-Squared the program gives you a lot of real-time flexibility.

One of the other very high-end performance products was Mandalla from Very Vivid. By connecting the computer, video camera/digitizer, graphics and music files a performer could interact with the computer.

The performer would stand in front of the camera and have his or her image digitized. The digitized image would be genlocked with graphics (like boxes, circles, drums, bells and other instruments) on the screen and the graphics would be linked to the music files. When the performer moved and the image came in contact with the graphics it would trigger the accompanying music file. If they touched the drum it would activate the drum sounds, if they touched a box the screen colors might change, etc. By dancing before the camera the computer images and sounds would be changed or triggered.

Another performance application is used in Joe Robbie stadium in Miami. During the 1989 Superbowl an Amiga was used to generate animations which were transferred to video tape and then displayed on the 40-foot RGB monitor called the Jumbotron.

There are also professional performers using the computer/video combination in their acts. Lilly Tomlin uses the Amiga to control lighting effects and the rock band Oingo Boingo does too. The band leader on the Pat Sajack show, Tom Scott, has used Amigas to control MIDI sequencers while an animated ballerina danced on a screen.

If you want to do any of these kinds of things keep in mind that when you get to this level you will need a lot of computer expertise and probably some custom hardware and software. Depending on who you are, what you want to do and how much you want to spend many companies will go a long way to work with you. I imagine if you are Frank Sinatra's stage manager and are interested in doing something with computer/video in a Las Vegas act there are a number of companies out there who would be more than happy to help.

# Post-production Houses

While you may not be able to do all the things that you see on network television with a desktop video set-up there are people that can. In most cities there are post-production houses that have all the equipment necessary to do professional editing and effects. Unfortunately, these post-production houses can be very expensive, sometimes costing hundreds of dollars per hour. If you have a lot of money to throw away then you can just set up an appointment, gather your tapes and have the house work with you to put your production together. If you really want to you can rent entire production studios, cameras, personnel, editors and everything else you can think of from start to finish. Just remember that even a 30 second commercial can cost tens of thousands of dollars to produce. There is a way to cut down on the costs of a post-production house that is used by many amateurs and even professionals.

A process called **off-line editing** is simply doing a rough edit yourself to generate a **working master.**

Rather than walking in to a post-production house with a handful of tapes and a script you can save quite a bit of money by doing as much of the editing and post-production work as possible with your home equipment. By making a working master, before you go to the post-production house, you will have a much better idea how things will look, how well they work together and you will have a tape that the post-production editor can refer to. This way, when you show up at the studio, you will already know how you want things and won't have to spend extra time deciding if things look right.

You can just bring in your raw footage and your working master but there are two things that professionals do to speed things up even more. (Even professionals don't want to waste a lot of time when it is costing them $250 an hour). The first thing that anyone can do no matter what equipment you have is to create a detailed **Edit Decision List** called an **EDL**. The EDL is like your script except that it includes exact information about the order, transitions, times and location of every single shot in the sequence as they will appear in the final production. An EDL usually has only a word or two description of the shots themselves (just enough description to confirm that you have the correct shot later).

There are three varieties of EDLs; control track, SMPTE and computer readable. Control track, just uses the tape counter to generate the in and out points in the EDL. Since most consumer VCRs have sloppy tape counters the numbers will be a little sloppy too. SMPTE uses much more exact numbers. A SMPTE time code track can be laid down on blank tape before you shoot, during shooting or afterwards. Basically, it is a digital counter where the numbers are recorded on part of the tape where it won't interfere with the picture or audio. The numbers can only be read by special equipment with one important exception. There is a special piece of equipment called a SMPTE Window Dub Generator. A window dub generator just burns (overlays) the SMPTE time code numbers on top of your video. A window dub has a box with the numbers in it, as if you superimposed a digital clock on top of your video. Window dubs are ideal for creating an EDL and for off-line editing. Computer readable EDLs are not the kind of EDLs you can create with a home computer. Many post-production houses use special computers to perform their operations. All the edits and transitions are first entered into the computer and when everything is ready the computer performs all the operations in one pass. If you have the right equipment (fairly expensive equipment at that) you can enter all the things you want the post-production house to do for you, bring your raw footage tapes and the computer readable EDL (on diskette, paper tape or punched cards) plug everything in and zip through the entire production.

After you have done all your pre-production and production work, get yourself a pot of coffee, a pad of paper, a pencil and lock yourself in your studio. If you have a SMPTE window dub generator you can make copies of your tapes before you get started on the EDL. Since you won't be using the window dub copies for your final editing the quality of the copies doesn't matter. Now you can go back, zero your tape counter (if you are doing a control track EDL), and watch all of your tapes. You will be making detailed notes of every single shot, particularly counter numbers (both in and out points) and shot durations. You want to log every single shot, even the ones that you don't intend to use in the final production. While you are doing this you can jot down which shots you want to use later. When you are finished you can go back and start your EDL.

The first time through you will want to follow your script as close as possible. This may change later if you decide that things don't look quite right. A traditional EDL has the following information; the edit number, source (which can be the reel or cassette number, character generator, etc.), mode (audio, video, both), transition type, effect duration (in frames), playback in, playback out, record in and record out numbers. If you want you can also add your brief shot description too. Of course your record in and record out numbers will have to be calculated since those numbers don't exist until after the editing is finished. In the desktop video world you might want to be a little flexible here and include things like computer effects even if they won't show up until the editing process.

If you are doing your own final editing then you can charge on ahead once you have your EDL completed. Even if you are planning to go to a post-production house for your final editing you might want to do a working master anyway to give you an idea of how everything will look. If you are doing a working master you should use the window dub copies (if you have them) or use copies of your original footage tapes (remember quality isn't important here and you don't want to risk your original material).

Once you have your EDL and working master you can set up your appointment with the post-production house. Be sure to tell them everything you have, what you plan to do, how many edits, what kinds of effects, what format your original footage is in, what format you want your final master to be in, what kind of SMPTE code you are using (there are a number of different formats) and any other details you can think of. This is the second time and money saving step that everyone should do when going to a post-production house.

They will have their own list of questions and should be able to give you an idea of the time and costs (although if you start changing things in the middle it can add a lot of time). While they should already have things set up before you get there it is not unusual for them to add set up time to your bill. The more you can get done ahead of time and the more you can clear up on the phone the better. Most post-production houses will do just about anything you want because, after all, you are the one paying for their time.

Going to a post-production house can be expensive but the results are usually much better than what you could do with home equipment. While you can expect better results if you start with better quality raw footage, most post-production houses will dub over to another format to do the editing and for generating the master tape. They will then dub down to whatever format you need. They should even have all the equipment necessary to make a clean dub up from VHS or 8mm (but don't expect miracles).

## Summary

There are a number of advanced techniques you can use in the desktop video environment but most of them are yet to be discovered. In most cases it is going to depend on the application and will be highly specialized. The trick to getting professional results is usually to start with professional equipment. As you start wanting or requiring higher quality you have to start learning more about the equipment that is available and how that equipment works. If you don't know what is possible or, at the other extreme, you overestimate your equipment's capabilities it is going to be difficult to get professional results.

If you don't have the money to spend then careful planning and a creative imagination will be your primary tools. We have all seen locally produced programs or commercials that look and sound 'cheap' but even these may have cost someone tens of thousands or even hundreds of thousands of dollars. You may want to include effects like those we see in commercials, broadcast TV or films but not all of them are possible with desktop video equipment. If you can focus on the things that you know you can do well with the equipment at hand then your productions will be much, much better in the end. Some of the most effective and clean productions ever produced are very simple and don't use hundreds of special effects. Compare high-tech horror films like Nightmare on Elm Street to low-tech horror films like Psycho.

In any of these areas there are books and magazines that can be invaluable. And there are people out there who can help you. Since desktop video is a new field that is growing every day the 'professionals' are people like you who have found out a few things by trial and error and most of them would be more than happy to help. You can try your local cable or broadcast station to find out if they are using Amigas already. If they are they might be able to give you some tips. If they are not using them yet they might be interested enough to help you out because they could eventually benefit from your efforts. Finally, if there is enough interest and support for this book I might even do another for the more advanced users out there.

# Chapter 13

## Conclusion

# Chapter 13
# Conclusion

While I was putting this book together I had the opportunity to talk to a number of people who were either doing desktop video themselves or were developing products for desktop video. The thing that struck me is the amount of interest in the subject and the amount of confusion.

When I was working in video full time, back in the mid-seventies, we had what was considered state-of-the-art cable TV equipment (at the time). But the equipment was expensive and by todays standards not very professional. We ran into a number of problems that were overcome mostly through extra work and ingenuity. When I left video for writing and then computers I lost track of the advances in video technology. In recent years computer technology and video technology have come together and I was excited about the idea of doing this book. I could get back into the video side of things and re-educate myself. I was pleasantly surprised by the advances in 'consumer level' video and a little surprised to find that the state of consumer video is about at the same level professional cable TV was when I left it. Many of the tricks and techniques that I have outlined in this book are the same things that I had to use back in the seventies. Of course we didn't have personal computers back then so many of these ideas required spending thousands of dollars on special equipment.

Another thing that surprised me was how serious the users and manufacturers feel about consumer level video. The kinds of specialized equipment that manufacturers used to make costing thousands of dollars are now available to the consumer for hundreds of dollars. A few years ago people might only spend a few hundred dollars on their home video equipment but now it is not uncommon to find people spending thousands for Hi8 camcorders, S-VHS equipment, projection TVs, surround-sound, etc . These people (you may be one of them) want to do more than just make video versions of home-movies. They want to add titles, graphics, animations and edit their videos. They are also interested in making money with their video equipment doing weddings, real estate tapes, legal video-depositions, advertising and other projects. These people are willing to spend extra money on special equipment to enhance their productions, to do editing, titling, etc., and manufacturers realize this.

The other thing I noticed right away was the confusion. As I said in the introduction, most people either understand the video side or the computer side but few people understand both. If you have read this far then you should now understand that neither field is terribly difficult. The problem has been (up until now) that you couldn't find all the information in one, nice neat package. I don't consider myself to be the world's foremost authority on computers or video. I had to rely on books and articles from a dozen different places. The problem was that the information is scattered and sometimes difficult to find. Of course, one of the reasons that you bought this book was to have someone else do all that research for you.

I know that there are things that I left out or skimmed over but I think we covered quite a bit of ground. I wanted to give you the basics and enough to get a good start in this relatively new field. I'm sure that I missed some products and may have made a few errors along the way but nobody is perfect.

The future of desktop video is unlimited. It is just getting started and I am sure that we will all be seeing some remarkable things. Already we are seeing the quality of consumer equipment getting better and better while the prices have been dropping. More and more people own camcorders and computers and more and more companies are coming out with hardware and software to combine the two. If you look in video magazines these days you will see articles and even columns devoted to desktop video. Private companies and schools are beginning to install studios and hire video people. Courses are being offered in computer graphics and even grade schools are teaching kids how to use video equipment. Video is being turned over (or taken over) by average people and the first results are impressive. Of all the video generated during the San Francisco earthquake of 1989 most of it was done by amateurs with camcorders not the networks. Home video is still in its early stages and desktop video even earlier.

I think it is going to be incredibly exciting to see what will happen in the next few years. No one can argue that video has changed the world and the way people think. Desktop video has that potential too. It is a bit frightening and also exhilarating at the same time. I know that I'm going to enjoy it and hope that we all will.

Guy Wright - November 1989

# Appendix A

## Related Products

# Appendix A
# Related Products

There are a few products that I didn't mention in the book that are related to desktop video but either didn't quite fit in or weren't ready at the time of this writing.

**MediaPhile and MediaProcessor**

Interactive Microsystems
80 Merrimack St.
P.O. Box 1446
Haverhill, MA 01830

The MediaPhile and MediaProcessor systems allow you to control video equipment directly from the computer. With hardware modifications done to just about any VCR (either done buy the user or installed by the company) you can get very accurate readings, control all the functions, create EDLs and perform editing all from the computer. In a way you are turning the computer into a kind of editor/controller. You can also use the system to create presentations or video data bases. Interactive Microsystems sells a complete line of Amiga hardware and software plus video equipment. One of the nice features of the system is its ability to learn and transmit infrared signals. Any equipment you have that can be controlled with an infrared device can be controlled by the computer. The system is very flexible and not overly expensive (below $500).

**Complete Systems**

RGB Video Creations
3944 Florida Blvd., Suite 102
Palm Beach Gardens, FL 33410

If you want to buy a desktop video system all in one shot then RGB offers a complete solution in one package. Their system includes S-VHS equipment, genlocking devices, fully equipped Amiga and software that lets you control everything. For about $10,000 you can have a complete S-VHS editing suite, tested, wired and ready to go.

## Vmachine

Digital Creations
2865 Sunrise Blvd., Suite 103
Rancho Cordova, CA 95742

Digital Creations, manufacturers of the SuperGEN genlock is nearly ready to bring out an entire line of video products including S-VHS genlock, frame buffer including digitizing and frame synchronization with built-in TBC (time base corrector) and paint box. The packages should let you do A/B roll editing with full cross dissolves, overlays, keying, mattes, flys and a full series of real time DVE (digital video effects or ADO) effects.

## Video Toaster

NewTek
115 W.Crane St.
Topeka, KS 66603

The Video Toaster, when it appears, will be a real time DVE device that will let you manipulate video images in a number of ways. You will be able to do flips, spins, shrinks, mosaics, mapping, polarizations, pixelizations and many other special effects. If it lives up to half of the things they claim it will do it should be a rather remarkable device.

## Photon Video

Microillusions
17408 Chatsworth St.
Granada Hills, CA 91344

Microillusions has a few products for the more advanced desktop video producer. Their transport controller works with single frame video recorders and controllers like the Lyon Lamb MINIVAS, VAS IV, Videomedia V-LAN, BCD 2500 or Pico Systems Animation Controller. The software works with Microillusion's Photon Cell Animator and other animation packages to create smooth single frame animations. Their EDLP is a program for creating and manipulating EDLs. Microillusions also sells a time code reader/generator called the TCRG-102 that features longitudinal/SMPTE/EBU time code, drop frame or non-drop frame time code, Color frame accurate, and it is compatible with their Music-X MIDI software.

## VIVA

MichTron
576 S. Telegraph
Pontiac, MI 48053

If you are interested in producing interactive video then you might look into VIVA (Visual Interfaced Video Authoring). It is an authoring language/interface that lets you customize an interactive video production. Mainly used with laser disks the system is fairly simple to use and lets you create what amounts to an IF...THEN list where the video is treated like a data base with random (as opposed to sequential) access. VIVA is similar to presentation programs giving you control over text, graphics, video, sound, color and animation. The system can control laser disk players, VCRs and other media devices (with the proper interfacing). It is ideal for education, training or even information kiosk type applications.

# Appendix B

# Magazines and Books

# Appendix B
# Magazines and Books

**Amiga Magazines**

Amazing Computing
PIM Publications, Inc.
Currant Rd.
P.O. Box 869
Fall River, MA 02722-0869

Amiga Plus
Antic Publishing Inc.
544 Second St.
San Francisco, CA 94107

Amiga Resource
Compute Publications, Inc.
324 West Wendover Ave.
Greensboro, NC 27408

AmigaWorld
IDG Communications, Inc.
80 Elm St.
Peterborough, NH 03458

Amigo Times
Sama Software, Inc.
5124 St. Laurent, Suite 100
Ville St. Catherine, QUE J0L 1E0  CANADA

INFO
Info Publications, Inc.
123 N. Linn St., Suite 2A
Iowa City, IA 52245

## Video Magazines

AV Video
Montage Publishing, Inc.
25550 Hawthorne Blvd., Sutie 314
Torrance, CA 90505

Camcorder Report
Miller Magazines, Inc.
2660 East Main St.
Ventura, CA 93003

Video Magazine
Reese Communications, Inc.
460 W. 34th St.
New York, NY 10001

VideoMaker
Videomaker, Inc.
1166 Lassen Ave.
Chico, CA 95926

## Books you might find useful

Today's Video
and
Video User's Handbook
by Peter Utz
Prentice Hall Press

Videotape Editing: A postproduction Primer
by Steven E. Browne
Focal Press

Desktop Video: A guide to personal and small business video production
by Austin H. Speed III
Harcourt Brace Jovanovich, Publishers

There are dozens of other books and magazines on video and computers. Check with your library, bookstore and newstand for other titles.

# Appendix C

# Amiga Video Sources

# Appendix C
# Amiga Video Sources

A-Squared Distributions Inc.
6114 LaSalle Ave.
Oakland, CA 94611
415-339-0339
LIVE! 2000

Access Technologies
P.O. Box 202197
Austin, TX. 78720
512-343-9564
3D fonts

Aegis Development Inc.
2115 Pico Blvd.
Santa Monica, CA. 90405
213-392-9972
Impact!, Aegis Animator, Aegis VideoTitler, Videoscape 3D, Lights! Camera! Action!

AlohaFonts
PO Box 2661
Fair Oaks, CA 95628
fonts

Antic Software
544 Second St.
San Francisco, CA 94107
415-957-0886
Zoetrope

ASDG
925 Stewart St.
Madison, WI 53713
608-273-6585
Professional Scan Lab

Broderbund Software Inc.
17 Paul Dr.
San Rafael, CA 94093-2101
800-521-6263
Fantavision

Byte-by-Byte
Aboretum Plaza II
9442 Capitol of Texas H-Way N., Suite 150
Austin, TX. 78759
512-343-4357
Animate-3D, Sculpt 4D

C Ltd.
723 East Skinner
Wichita, KS 67211
316-276-6322
C-View, Han-D-Scan

Charles Voner Designs
61 Clewley Rd.
Medford, MA 02155
617-396-8354
clip art

Commodore Business Machines
1200 Wilson Dr.
West Chester, PA 19380
215-436-4200
Various Amiga products

Communications Specialties Inc.
6090 Jericho Turnpike
Commack, NY 11725
516-499-0907
Gen/One

Computer Arts
PO Box 529
Opp, AL 36467
205-493-6312
clip art maps

Creative Microsystems, Inc.
10110 S.W. Nimbus #B1
Portland, OR. 97223
503-691-2552
V-I 2000

CV-Designs
61 Clewley Rd.
Medford, MA 02155
617-396-8354
Video Visions

Digital Creations
2865 Sunrise Blvd. Suite 103
Rancho Cordova, CA 95742
916-344-4825
SuperGen 2000S, V Machine, Living Color

Eagle Tree Software
P.O. Box 164
Hopewell, VA 23860
804-452-0623
Butcher

Elan Designs
P.O. Box 31725
San Francisco, CA 94131
415-359-7212
Invision, Elan Performer

Electronic Arts
1820 Gateway Dr.
San Mateo, CA 94404
800-245-4525
Delux-Paint III, Delux-Video, Delux-Productions, Delux Phot Lab

Free Spirit Software
P.O. Box 128
Kutztown, PA 19530
215-683-5609
Media Line Animation Backgrounds

Gold Disk
2175 Dunwin Dr. Unit 6
Mississauga, ONT CANADA L5L 1X2
MovieSetter

Hash Enterprises
2800 E. Evergreen Bld.
Vancouver, WA. 98661
206-693-7443
Animation: series (Apprentice, Effects, Flipper, Jr., Stand)

Impulse, Inc.
6870 Shingle Creek Parkway #112
Minneapolis, MN 55430
612-566-0221
Impulse Video Digitizer, Turbo Silver

InnoVision Technology
P.O. Box 743
Hayward, CA 94543
415-638-8432
Video Effects 3D, Broadcast Titler

InterActive Softworks Inc.
2521 South Vista Way Suite 254
Carlsbad, CA 92008
fonts

Interactive Microsystems
PO Box 1446
Haverhill, MA 01831
508-372-0400
MediaBase and MediaProcessor

JMH Software
7200 Hemlock Lane
Maple Grove MN 55369
612-424-5464
JMH Easy Titler

Kara Computer Graphics
6365 Green Valley Circle, No. 317
Culver City, CA 90230
213-670-0493
fonts

Magni Systems Inc.
9500 SW Gemini Dr.
Beaverton, OR. 97005
503-626-8400
Magni 4004

Meridian
9361 W. Brittany Ave
Littleton, CO 80123
303-979-4140
The Demonstrator

MichTron
576 S. Telegraph
Pontiac, MI 48053
313-334-5700
VIVA

Micro Magic
261 Hamilton Ave. #320 C.
Palo Alto, CA 94301
415-327-9107
Forms in Flight

Microillusions
17408 Chatsworth St
Granada Hills, CA. 91344
818-360-3715
Photon Video, Photon Video Edit 3D, Photon Video Render 3D, Photon Paint

Mimetics Corp.
P.O. Box 1560
Cupertino, CA 95015
408-741-0117
AmiGen, Frame Buffer, Frame Capture, 3-Demon, Sound Sampler

Mindware International
110 Dunlop W. Box 22158
Barrie, ONT CANADA L4M 5R3
705-737-5998
Pageflipper Plus FX

New Horizons Software
PO Box 43167
Austin, TX 78745
512-328-6650
fonts

NewTek
115 W. Crane St
Topeka, KS 66603
913-354-1146
Digi-View Gold w/camera and table, DigiDroid

PAR Software
Distributed by Brown-Wagh Publishing
16795 Lark Ave. Suite 210
Los Gatos, CA 94107
408-395-3838
Express Paint

Progressive Peripherals & Software
464 Kalamath St.
Denver, CO 80204
303-825-4144
Frame Grabber, ProGEN, PIXmate

PVS Publishing
3800 Botticelli
Suite 40
Lake Oswego, OR 97035

Pro Video CGI, Pro Video Plus

RGB Video Creations
3944 Florida Blvd. Suite 102
Palm Beach Gardens, FL 33410
407-622-0138

Right Answers Group
Box 3699
Torrance, CA 90510
213-325-1311
Director

Shereff Systems, Inc.
15075 SW Koll Pkwy, Suite G
Beaverton, OR. 97006
503-626-2022
JDK ImagesProVideo series

Silver Software
77 Mead St.
Bridgeport, CT 06610
203-366-7775
Fractal Music, DNA Music, Protein Music

Software Sensations
1441 South Robertson Blvd.
Los Angeles, CA. 90035
213-277-8272
S-View

Sparta Inc.
23041 de la Carlotta. Suite 400
Laguna Hills, CA 92653
714-583-2394
Lights!, Camera!, Action!, Animagic, Video Effects 3D

Sunrize Industries
3801 Old College Rd.
Bryan, TX 77801
409-846-1311
Color Splitter

Syndesis
N9353 Benson Rd.
Brooklyn, WI 53521
608-455-1422
InterChange

The Right Answers Group
P.O. Box 3699
Torrance, CA 90570
213-325-1311
The Director

VidTech Intl. Inc.
2822 NW 79th Ave.
Miami, FL. 33122
800-727-3361
Scanlock

Vivid Produce
PO Box 127 Stn B.
Toronto, ONT Canada M5T 2T3
416-686-7850
Interactor

Zuma Group
6733 North Black Canyon Hwy.
Phoenix, AZ 85015
602-246-4238
TV*Show, TV*Text Pro

# Index

| | | | |
|---|---|---|---|
| 8mm HD | 19 | Beta | 22 |
| 12-1 zoom lenses | 34 | Betacam | 18 |
| 68020 | 45 | Betacam-SP | 18 |
| 68030 | 45 | bi-directional | 13 |
| | | Bidirectional mikes | 143 |
| AB dissolve | 79 | black and white | 164, 165, 170 |
| AB effects | 79 | black and white video camera | 183 |
| Accelerator Board | 191 | black burst | 52 |
| Adjustable zoom speed | 34 | blit | 131 |
| ADO | 65, 113 | BNC | 52 |
| after-the-fact time coding units | 26 | bold | 125 |
| AI (artificial intelligence) | 149 | Boom mikes | 143 |
| All-in-one programs | 115 | Broadcast Titler | 55, 132 |
| Amiga 500 | 43, 181 | brush animation | 112 |
| Amiga 1000 | 43 | built-in character generators | 33 |
| Amiga 2000 | 43, 186 | burn in | 66 |
| Amiga 2000HD | 43, 191 | Butcher | 94 |
| Amiga Fonts | 123 | | |
| ANIM file | 88 | Cables | 145 |
| ANIM format | 75 | CAD | 73 |
| Animation | 171, 204 | Camcorder | 5, 180, 185, 190 |
| Animation mode | 35 | camera angles | 156 |
| Animation Program | 183, 188, 192 | capstan servo mechanism | 47 |
| Animation Software | 87 | Carbon mikes | 142 |
| animation backgrounds | 163 | Cardioid | 13 |
| animation packages | 9 | Cardioid mikes | 143 |
| animation/time lapse | 75 | Carrying Case | 38 |
| Anti-ailiasing | 61, 92, 126 | cast shadow | 125 |
| Apprentice | 107 | Cel Animator | 114 |
| assemble edits | 11 | Cell Animation | 87, 110 |
| Audio Dub | 33, 36, 186, 210 | cellulose film | 110 |
| Audio Mixer | 37, 144, 191 | center justified | 126 |
| audio | 5, 209 | Character Generator | 8, 34, 121 |
| audio digitizer | 13, 141, 187 | Chroma | 5 |
| audio dub capabilities | 141, 181 | Chroma Key | 9 |
| | | Chrominance | 5 |
| Back light | 34, 37 | clip-art disks | 121 |
| BACKUP COPY | 55 | Close Ups | 159 |
| Batteries | 37 | CMI VI-500 | 46 |

| | | | |
|---|---|---|---|
| Color | 124 | Digi-View Gold 3.0 | 63 |
| Color averaging | 61 | Digitizer | 59, 176, 183 |
| Color Corrector | 11 | Digitizing and frame grabbing | 203 |
| Color fonts | 124 | digitizing process | 67 |
| Color Mis-adjustment | 169, 170 | Directional and Unidirectional mikes | 143 |
| Color viewfinders | 35 | display programs | 121 |
| color bar charts | 95 | Dissolve | 46, 128, 162, 166 |
| color bar generator | 95 | dissolve rates | 54 |
| color cycling | 85, 87, 92, 109, 168 | Distortion | 169 |
| color splitter | 63, 64, 183 | Dithering | 61, 92 |
| color zero | 97 | down-stream | 7 |
| Composit | 5 | DPI | 62 |
| Composit Video | 5 | dub | 11 |
| Computer Overlays | 208 | DVE | 169, 170 |
| Computer to Tape Transfers | 207 | Dynamic | 13 |
| computer readable | 222 | Dynamic mikes | 142 |
| Computers | 59 | | |
| Condenser | 13 | ED Beta | 19, 25 |
| Condenser mikes | 142 | Edit Decision List | 222 |
| Consumer Video | 19 | Editing | 36, 160 |
| Control L | 12 | Editor | 114 |
| Control S | 12 | Editor/Controller | 12, 33, 186, 190 |
| control track | 6, 47, 222 | EDL | 209, 222 |
| control track EDL | 223 | Electret condenser mikes | 142 |
| counter | 181 | electronic fluctuations | 22 |
| Crawl | 128 | Encoder | 5, 45, 181 |
| Created Sounds | 176 | execute | 33 |
| Crystal | 13 | Experimental Non-program Information | 6 |
| Crystal and Ceramic mikes | 142 | External control jacks | 35 |
| cue points | 210 | external control | 30 |
| custom fonts | 124 | external control ports | 181 |
| Cut or Bang | 127 | external key in | 52 |
| Cuts | 156 | external video | 52 |
| Cycling | 125 | Extra Objects | 167, 171, 172 |
| | | Extruding | 125 |
| D2 | 18 | | |
| Deluxe Music Construction Set | 148 | F/X | 155 |
| Deluxe Paint III | 92 | Fade | 128, 157, 162 |
| Deluxe PhotoLab | 94 | Fade controls | 34 |
| Deluxe Productions | 130 | Fantavision | 116 |
| demonstration | 220 | FCC | 4 |
| Desktop Video | 3 | fields | 6 |
| desktop video setups | 179 | Fill light | 37 |
| Digi-Paint | 3, 63, 93 | | |
| Digi-View | 63 | | |

| | |
|---|---:|
| filled shadow | 125 |
| filter wheel | 64 |
| Filters | 37, 159, 169 |
| Flashing | 125 |
| Flip | 128 |
| Flipper | 114 |
| flipping | 87 |
| flush justification | 126 |
| Flying Erase Heads | 11, 23, 27, 34, 35, 181 |
| focus | 156 |
| Foley Sounds | 175 |
| font | 123 |
| font packages | 86, 121 |
| foreshadowing | 156 |
| four or more heads | 181 |
| Frame Buffer | 71, 72, 191 |
| Frame Grabber | 9, 61, 71, 72, 187 |
| Frame Store | 47 |
| frame | 6 |
| frame store units | 72 |
| Framebuffer w/Frame capture | 77 |
| Framegrabber | 74 |
| Free | 126 |
| Freeze Frame F/X | 158 |
| freeze frame | 36, 165 |
| Gen/One | 51 |
| generation | 11 |
| Genlock | 3, 8, 47, 91, 97, 191 |
| Genlocking Device | 183, 187 |
| glitch | 11, 27, 161 |
| GrabANIM | 131 |
| grabbing images | 80 |
| graphic interface | 44 |
| grid snap | 126 |
| halfbrite | 92 |
| HAM mode | 92, 99 |
| Hand Controlled Animation | 109 |
| Hand drawn materials | 62 |
| Hand-held | 143 |
| Hard Disk Drive | 187 |
| Hawkeye | 18 |
| Hi8 | 24 |
| high impedance | 13, 143 |
| high speed shutters | 159 |
| high-end consumer | 24 |
| highest resolution | 60 |
| Horizontal | 5 |
| hot keys | 54 |
| hot spots | 66 |
| HQ circuitry | 22, 36, 181 |
| Hybrid Animation | 113 |
| IFF | 60 |
| illegal colors | 97 |
| Image Enhancement | 170 |
| Image Enhancer | 11 |
| Image Enhancer/Color Corrector | 186, 190 |
| Image Master | 50 |
| Image Processing Program | 184, 188, 192 |
| Image processing | 9, 61, 170, 206 |
| Image processors | 86, 94, 98, 99 |
| image processing software | 86 |
| in-camera time coding | 26 |
| insert edits | 11 |
| Instant Music | 147 |
| Interactive demos | 220 |
| InterChange | 107 |
| Interfield Jitter | 73 |
| Interlaced mode | 6, 91 |
| IRE | 97 |
| Italics | 125 |
| jittering | 73 |
| jog-shuttle controls | 36, 186 |
| Justification | 126 |
| Kerning | 126 |
| Key light | 36 |
| Key out | 52 |
| Lavalier or Clip on mikes | 143 |
| Left justified | 126 |
| Lenses and Filters | 185, 190 |
| lighting | 14, 36, 66 |
| Lights, Camera, Action | 131 |

| | |
|---|---|
| Line Background | 127 |
| Line Transitions | 127 |
| line art' mode | 163 |
| line conditioner | 217 |
| linear time | 181 |
| LIVE! | 74 |
| Lockable palettes | 92 |
| Low impedance | 13, 143 |
| Low Lux numbers | 34 |
| low-end professional | 24 |
| Luma. | 5 |
| Luminance | 5 |
| LUX numbers | 14 |
| | |
| M-2 | 18 |
| Macro lenses | 159 |
| Macro-zoom lenses | 34 |
| Magni4004 | 50 |
| Mapping | 55, 169, 170 |
| Masks | 169 |
| master deck | 11 |
| matte painting | 167 |
| Memory Board | 186, 191 |
| Memory expansion | 45 |
| Metamorphic | 87 |
| metamorphic animation | 112 |
| Microillusions Transport Controller | 114 |
| Microphone | 37, 142, 184, 185, 190 |
| MIDI | 147 |
| MIDI device | 88 |
| MIDI Instrument | 187, 191 |
| MIDI Interface | 187, 191 |
| MIDI software | 149 |
| Mis-adjustment | 160, 163, 164, 166 |
| Miscellaneous Items | 182 |
| Miscellaneous Software | 89 |
| Mix and Match philosophy | 26 |
| mixer | 13, 144, 183 |
| Modeling or Kicker light | 37 |
| Montage work | 62 |
| montage | 164, 165 |
| mouse control | 44 |

| | |
|---|---|
| MovieSetter | 115 |
| Multiple heads | 35 |
| Multiple Images | 164 |
| Music Program | 88, 184, 188, 192 |
| music | 139, 174 |
| | |
| Negative/Positive special effects | 35 |
| Newspaper Photographs | 164, 165 |
| non-interlaced modes | 6 |
| non-interlaced pictures | 6 |
| non-overscanned graphic | 91 |
| non-real-time digitizers | 61 |
| noncomposit | 5 |
| note editing packages | 148 |
| NTSC | 4 |
| | |
| off-line editing | 222 |
| Old Style Photographs | 164, 165 |
| Omni-directional | 13 |
| Omnidirectional mikes | 143 |
| On screen programming | 36 |
| One inch decks | 18 |
| Outlining | 125, 196 |
| Over dubbing | 161 |
| overlay graphics | 52 |
| Overlay1 | 52 |
| Overlay2 | 52 |
| Overlays | 161 |
| oversaturated | 98 |
| overscan | 6 |
| overscanned | 91 |
| | |
| Page Flipping Program | 113, 188, 192 |
| Page Render3D | 106 |
| page flipping | 110 |
| page flipping software | 90 |
| PageFlipper Plus F/X | 114 |
| Paint Program | 182, 188, 192 |
| Paint Program, rendering and ray traced Graphics | 203 |
| Paint programs | 9, 85 |
| PAL | 4 |
| Parabolic mikes | 143 |
| PerfectSound Digitizer | 149 |

| | | | |
|---|---|---|---|
| perform edit | 33 | Rendering Program | 103, 188, 192 |
| Photon Paint 2.0 | 93 | RF (Radio Frequency) | 6 |
| picture quality | 20 | RGB | 5 |
| pixel | 60 | right justified | 126 |
| Pixelization | 164, 165 | Roll or Scroll | 128 |
| PIXmate | 95 | Rotoscoping | 173 |
| Planning | 141, 194 | | |
| point size | 123 | S-VHS | 19, 24 |
| Pointers | 168 | S-VHS-C | 19, 24 |
| poly fonts | 133 | saturation | 54 |
| Positioning | 126 | Say Speech Synthesizer | 146 |
| post-production houses | 221 | Scanlock | 51 |
| pre-production | 194 | scanned image | 62 |
| pre-roll | 12 | Scanner | 9, 59, 62 |
| pre-roll edits | 33 | Screen manipulation | 113 |
| Presentation (Display Only) Programs | | Screen Transitions | 127 |
| | 122 | screen background | 127 |
| Presentation Program | 188, 192 | scripting | 198 |
| Presentation software | 88, 121 | Sculpt-Animate 4D | 105 |
| presentation | 218 | Sculpt-Animate 4D Jr | 105 |
| presentation scripting process | 206 | SECAM | 4 |
| Pressure Zone Microphones | 143 | SEG | 88 |
| Preview | 52 | self running demo | 220 |
| print to video | 163 | Set light | 37 |
| Printer | 191 | Shadowing | 125 |
| Pro Video Gold | 132 | Shotgun or Zoom mikes | 143 |
| Proc Amp | 10, 186 | Shutter speed | 35 |
| Professional ScanLab (PSL) | 63 | Single Frame animation | 10 |
| ProGEN | 49 | single frame advance | 181 |
| pseudo-extrusion | 125 | slave deck | 11 |
| publishing | 73 | Slide Off | 128 |
| Push | 128 | Slide On | 128 |
| | | slider bars | 98 |
| Quad decks | 17 | slips | 21 |
| quad tape | 72 | slow motion | 181 |
| Quartercam | 18 | SMPTE | 222 |
| | | SMPTE Time Code Generator/Reader | |
| Radio microphones | 143 | | 190 |
| RAM animation | 10 | SMPTE time code track | 222 |
| Ray Tracing | 9, 45 | SMPTE time coding | 12 |
| Ray Tracing packages | 87, 104 | SMPTE Window Dub Generator | |
| Re-Drawing | 171 | | 222 |
| real-time digitizers | 61, 72 | Sonix | 148 |
| Recam (type M) | 18 | Sony 5 Pin | 12 |
| Rendering | 9 | Sound System | 183, 191 |

255

| | | | |
|---|---|---|---|
| sound digitizers | 176 | The Director | 131 |
| sound effect | 139 | Time Base Corrector (TBC) | 10 |
| sound sampler | 141, 147, 149, 176 | time lapse | 33, 34, 158 |
| Sounds | 174 | Titler | 133 |
| Soundtrack | 150 | Titling Program | 184, 188 |
| SP (Standard Play) mode | 22 | Titling Program Screen/Text | 192 |
| Special Effects Generator (SEG) | 8 | Titling Program Screen/Text/Display | 192 |
| Speed Changes | 161 | titling | 121, 163, 205 |
| spline patches | 107 | titling package | 86 |
| Stand mikes | 143 | Touch Ups | 166 |
| standards | 4 | "track" | 27 |
| start tape deck | 210 | traditional EDL | 223 |
| stationary erase head | 27 | Transforming | 87 |
| Stereo | 36 | Tripods | 36 |
| Stop Watch | 38 | Tumble | 128 |
| Straight Animation | 172 | Turbo Silver 3.0 | 106 |
| Straight Titling | 174 | TV*SHOW | 130 |
| strobe | 159 | TV*TEXT PRO | 129 |
| Structured drawing packages | 87 | Tweening | 88 |
| Structured Drawing. | 86, 103 | Tweening Animation | 111 |
| super-fast shutter speeds | 33 | typical assemble edit | 30 |
| Superbeta | 19, 23 | | |
| SuperGEN | 49 | U-Matic decks | 18 |
| Switcher | 8 | U-SP deck | 18 |
| Switcher/Fader | 8, 191 | underline | 125 |
| Sync Generator | 8, 47, 190 | Uni-directional | 13 |
| Sync Signal | 5 | up-stream | 7 |
| sync | 5 | | |
| | | VCR | 181, 186, 190 |
| Tape | 38 | VD-1 | 76 |
| Tape Decks | 144 | Vector Scope | 10, 190 |
| Tape tips | 38 | Vertical Interval | 5 |
| tape in | 210 | Vertical Interval Reference Signal | 5 |
| tape out | 210 | Vertical Interval Test Signal | 5 |
| tape vibrations | 22 | Vertical Sync Pulse | 5 |
| TBC (time base corrector) | 47 | vertical distortion | 22 |
| TBC | 21 | vertical synchronizing signals | 5 |
| Telephoto mikes | 143 | VHS | 21 |
| Teletype | 128 | VHS-C | 19, 22 |
| Text/Screen generating programs | 122 | VHS-HQ | 19 |
| text/screen only programs | 121 | VI-2000 | 46 |
| Text/Screen/ Display Programs | 121, 123 | Video Cassette Player | 7 |
| | | Video Cassette Recorder | 7 |
| texture mapping | 106 | Video Digitizer | 9 |

| | |
|---|---:|
| Video Effects 3D | 129 |
| Video reference in | 52 |
| Video signals | 91 |
| Video Tape Player | 7 |
| Video Tape Recorder | 7 |
| Video tape editing | 11, 209 |
| Video thru | 52 |
| video | 5, 91 |
| video camera | 4 |
| video heads | 12 |
| video in | 5, 52 |
| video out | 5, 52 |
| video picture | 27 |
| video player | 7 |
| video recorder | 7 |
| video signal | 5 |
| video software | 85 |
| VideoTitler | 133 |
| voice over | 13 |
| voice overs | 139, 159, 175 |
| | |
| Waveform Monitor | 10, 97, 190 |
| white level | 158 |
| window dub | 222 |
| Wipe | 46, 127 |
| working master | 222 |
| | |
| "Zero Frame Editing" | 27 |
| Zoetrope | 115 |
| Zooms | 157 |

# Books for the AMIGA

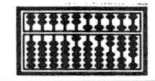

## Amiga for Beginners

**Amiga For Beginners-** the first volume in our Amiga series, introduces you to Intuition (Amiga's graphic interface), the mouse, windows, the CLI, and Amiga BASIC and explains every practical aspect of the Amiga in plain English. The glossary, "first-aid" appendix, icon appendix and technical appendix are invaluable to the beginner.

Topics include:

- Unpacking and connecting the Amiga components
- Starting up your Amiga
- Customizing the Workbench
- Exploring the Extras Disk
- Taking your first steps in the AmigaBASIC programming language
- AmigaDOS functions
- Using the CLI to perform 'housekeeping' chores
- First Aid, Keyword, Technical appendixes
- Complete set-up instructions
- Backing up important diskettes
- Setting Preferences
- Creating your own icons

**No Optional Disk Available**

**Volume 1   Suggested retail price    $16.95  ISBN 1-55755-021-2**

## AmigaBASIC: Inside & Out

**AmigaBASIC- Inside and Out-** THE definitive step-by-step guide to programming the Amiga in BASIC. Every AmigaBASIC command is fully described and detailed. Topics include charts, windows, pull down menus, files, mouse and speech commands.

Features:

- Loaded with real working programs
- Video titling for high quality object animation
- Windows
- Pull-down menus
- Moused commands
- Statistics
- Sequential and random files
- Exciting graphics demonstrations
- Powerful database
- Charting application for creating detailed pie charts and bar graphs
- Speech utility for remarkable human voice syntheses demonstrations
- Synthesizer program to create custom sound effects and music.

**Volume 2   Suggested retail price    $24.95  ISBN 0-916439-87-9**

| Optional Diskette | $14.95 | #612 |

Save Time and Money!-Optional program disks are available for all our Amiga reference books (except Amiga for Beginners and AmigasDOS Quick Reference). Programs listed in the book are on each respective disk and saves countless hours of typing! $14.95

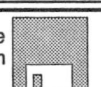

# Books for the AMIGA

## Amiga 3-D Graphics Programming in BASIC

Shows you how to use the powerful graphics capabilities of the Amiga. Details the techniques and algorithm for writing three-dimensional graphics programs: ray tracing in all resolutions, light sources and shading, saving graphics in IFF format and more.

Topics include:

- Basics of ray tracing
- Using an object editor to enter three-dimensional objects
- Material editor for setting up materials
- Automatic computation in different resolutions
- Using any Amiga resolution (low-res, high-res, interface, HAM)
- Different light sources and any active pixel
- Save graphics in IFF format for later recall into any IFF compatible drawing program
- Mathematical basics for the non-mathematician

**Volume 3   Suggested retail price   $19.95   ISBN 1-55755-044-1**

| Optional Diskette | $14.95 | #677 |

## Amiga Machine Language

**Amiga Machine Language** introduces you to 68000 machine language programming presented in clear, easy to understand terms. If you're a beginner, the introduction eases you into programming right away. If you're an advance programmer, you'll discover the hidden powers of your Amiga. Learn how to access the hardware registers, use the Amiga libraries, create gadgets, work with Intuition and much more.

- 68000 address modes and instruction set
- Accessing RAM, operating system and multitasking capabilities
- Details the powerful Amiga libraries for using AmigaDOS
- Speech and sound facilities from machine language
- Simple number base conversions
- Text input and output
- Checking for special keys
- Opening CON: RAW: SER: and PRT: devices
- New directory program that doesn't access the CLI
- Menu programming explained
- Complete Intuition demonstration program including Proportional, Boolean and String gadgets.

**Volume 4   Suggested retail price   $19.95   ISBN 1-55755-025-5**

| Optional Diskette | $14.95 | #662 |

Save Time and Money!-Optional program disks are available for all our Amiga reference books (except Amiga for Beginners and AmigasDOS Quick Reference). Programs listed in the book are on each respective disk and saves countless hours of typing! $14.95

# One Good Book deserves another
## and another, and another, and a...

**Amiga C for Advanced Programmers**
-contains a wealth of information from the pros: how compilers, assemblers and linkers work, designing and programming user friendly interfaces using Intuition, combining assembly language and C codes, and more. Includes complete source code for text editor.
ISBN 1-55755-046-8  400 pp  **$24.95**

**Amiga C for Beginners**
-an introduction to learning the popular C language. Explains the language elements using examples specifically geared to the Amiga. Describes C library routines, how the compiler works and more.
ISBN 1-55755-045-X  280 pp  **$19.95**

**Amiga 3-D Graphic Programming in BASIC**
-shows you how to use the powerful graphic capabilities of the Amiga. Details the techniques and algorithms for writing three-dimensional graphic programs: ray tracing in all resolutions, light sources and shading, saving graphics in IFF format and more.
ISBN 1-55755-044-1  300 pp  **$19.95**

**Amiga Disk Drives Inside & Out**
-is the most in-depth reference available covering the Amiga's disk drives. Learn how to speed up data transfer, how copy protection works, computer viruses, Workbench and the CLI DOS functions, loading, saving, sequential and random file organization, more.
ISBN 1-55755-042-5  360 pp  **$29.95**

**Amiga For Beginners** *
-the first volume in our Amiga series, introduces you to Intuition (Amiga's graphic interface), the mouse, windows, the CLI, and Amiga BASIC and explains every practical aspect of the Amiga in plain English.
ISBN 1-55755-021-2  184 pp  **$16.95**
*Includes Workbench 1.3 Info*

**AmigaBASIC Inside & Out**
-THE definitive step-by-step guide to programming the Amiga in BASIC. Every AmigaBASIC command is fully described and detailed. Topics include charts, windows, pulldown menus, files, mouse and speech commands.
ISBN 0-916439-87-9  554 pp  **$24.95**
*Includes Workbench 1.3 Info*

**Amiga Tricks & Tips**
-follows our tradition of other Tricks and Tips books for CBM users. Presents dozens of tips on accessing libraries from BASIC, custom character sets, AmigaDOS, sound, important 68000 memory locations, and much more!
ISBN 0-916439-88-7  348 pp  **$19.95**

**AmigaDOS Inside & Out**
-covers the insides of AmigaDOS from the internal design up to practical applications. Includes detailed reference section, tasks and handling, DOS editors ED and EDIT, how to create and use script files, multitasking, and much more.
ISBN 1-55755-041-7  280 pp  **$19.95**
*Includes Workbench 1.3 Info*

**Amiga Machine Language**
-is a comprehensive introduction to 68000 assembler machine language programming and is THE practical guide for learning to program the Amiga in ultra-fast ML. Also covers 68000 microprocessor address modes and architecture, speech and sound from ML and much more.
ISBN 1-55755-025-5  264 pp  **$19.95**

**Amiga System Programmer's Guide**
-comprehensive guide to what goes on inside the Amiga in a single volume. Only a few of the many subjects covered include the EXEC structure, I/O requests, interrupts and resource management, multitasking functions and much, much more.
ISBN 1-55755-034-4  442 pp  **$34.95**

**AmigaDOS Quick Reference** *
-an easy-to-use reference tool for beginners and advanced programmers alike. You can quickly find commands for your Amiga by using the three handy indexes designed with the user in mind. All commands are in alphabetical order for easy reference.  *Includes Workbench 1.3 Info*
ISBN 1-55755-049-2  128 pp  **$14.95**

**Computer Viruses: a high-tech disease** *
-describes what a computer virus is, how viruses work, viruses and batch files, protecting your computer, designing virus proof systems and more.
ISBN 1-55755-043-3  292 pp  **$18.95**
*"Probably the best and most current book...a bevy of preventive measures"*
*PC Week 11-21-88*

**Save Time and Money!**-Optional program disks are available for many of our Amiga reference books. All programs listed in the books are on each respective disk and will save you countless hours of typing!  **$14.95**
(* Optional Diskette Not Available for these Titles)

## Abacus
Dept. AA, 5370 52nd Street SE
Grand Rapids, MI 49512
**(616) 698-0330**

See your local Dealer or Call Toll Free **1-800-451-4319**

Add $4.00 Shipping and Handling per Order
Foreign add $12.00 per item

# Books for the AMIGA

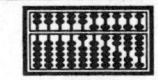

## AmigaDOS Quick Reference Guide

AmigaDos Quick Reference Guide is an easy-to-use reference tool for beginners and advanced programmers alike. You can quickly find commands for your Amiga by using the three handy indexes designed with the user in mind. All commands are in alphabetical order for easy reference. The most useful information you need fast can be found- including:

- All AmigaDOS commands described, including Workbench 1.3
- Command syntax and arguments described with examples
- CLI shortcuts
- CTRL sequences
- ESCape sequences
- Amiga ASCII table
- Guru Meditation Codes
- Error messages with their corresponding numbers

<u>Three</u> indexes for quick information at your fingertips! The AmigaDOS Quick Reference Guide is an indispensable tool you'll want to keep close to your Amiga.

**Suggested retail price US $9.95    ISBN 155755-049-2**

## Abacus Amiga Books

| | | | |
|---|---|---|---|
| Vol. 1 | Amiga for Beginners | 1-55755-021-2 | $16.95 |
| Vol. 2 | AmigaBASIC Inside & Out | 0-916439-87-9 | $24.95 |
| Vol. 3 | Amiga 3D Graphic Programming in BASIC | 1-55755-044-1 | $19.95 |
| Vol. 4 | Amiga Machine Language | 1-55755-025-5 | $19.95 |
| Vol. 5 | Amiga Tricks & Tips | 0-916439-88-7 | $19.95 |
| Vol. 6 | Amiga System Programmers Guide | 1-55755-034-4 | $34.95 |
| Vol. 7 | Advanced System Programmers Guide | 1-55755-047-6 | $34.95 |
| Vol. 8 | AmigaDOS Inside & Out | 1-55755-041-7 | $19.95 |
| Vol. 9 | Amiga Disk Drives Inside & Out | 1-55755-042-5 | $29.95 |
| Vol. 10 | Amiga C for Beginners | 1-55755-045-X | $19.95 |
| Vol. 11 | Amiga C for Advanced Programmers | 1-55755-046-8 | $34.95 |
| Vol. 12 | More Tricks & Tips for the Amiga | 1-55755-051-4 | $19.95 |
| Vol. 13 | Amiga Graphics Inside & Out | 1-55755-052-2 | $34.95 |
| Vol. 14 | Amiga Desktop Video Guide | 1-55755-057-3 | $19.95 |
| AmigaDOS Quick Reference Guide | | 1-55755-049-2 | $ 9.95 |

# Abacus Products for *Amiga* computers

## Professional DataRetrieve

### The Professional Level Database Management System

**Professional DataRetrieve**, for the Amiga 500/1000/2000, is a friendly easy-to-operate professional level data management package with the features most wanted in a relational data base system.

**Professional DataRetrieve** has complete relational data mangagement capabilities. Define relationships between different files (one to one, one to many, many to many). Change relations without file reorganization.

**Professional DataRetrieve** includes an extensive programming laguage which includes more than 200 BASIC-like commands and functions and integrated program editor. Design custom user interfaces with pulldown menus, icon selection, window activation and more.

**Professional DataRetrieve** can perform calculations and searches using complex mathematical comparisons using over 80 functions and constants.

**Professional DataRetrieve** is a friendly, easy to operate programmable RELATIONAL data base system. **PDR** includes PROFIL, a programming language similar to BASIC. You can open and edit up to 8 files simultaneously and the size of your data fields, records and files are limited only by your memory and disk storage. You have complete interrelation between files which can include IFF graphics. NOT COPY PROTECTED. ISBN 1-55755-048-4

MORE features of **Professional DataRetrieve**

Easily import data from other databases....file compatible with standard **DataRetrieve**....supports multitasking...design your own custom forms with the completely integrated printer mask editor....includes PROFIL programming language that allows the programmer to custom tailor his database requirements...

MORE features of PROFIL include:

Open Amiga devices including the console, printer, serial and the CLI.
Create your own programmable requestors
Complete error trapping.
Built-in compiler and much, much more.

**Suggested retail price:**                $295.00

### Features

- Up to 8 files can be edited simultaneously
- Maximum size of a data field 32,000 characters (text fields only)
- Maximum number of data fields limited by RAM
- Maximum record size of 64,000 characters
- Maximum number of records disk dependent (2,000,000,000 maximum)
- Up to 80 index fields per file
- Up to 6 field types - Text, Date, Time, Numeric, IFF, Choice
- Unlimited number of searches and subrange criteria
- Integrated list editor and full-page printer mask editor
- Index accuracy selectable from 1-999 characters
- Multiple file masks on-screen
- Easily create/edit on-screen masks for one or many files
- User-programmable pulldown menus
- Operate the program from the mouse or the key board
- Calculation fields, Data Fields IFF Graphics supported
- Mass-storage-oriented file organization
- Not Copy Protected, NO DONGLE; can be installed on your hard drive